CULTIVATING WORKERS

Cultivating Workers
Peasants and Capitalism
in a Sudanese Village

View of Wad al Abbas, 1988

VICTORIA BERNAL

NEW YORK
COLUMBIA UNIVERSITY PRESS

Columbia University Press
New York Oxford
Copyright © 1991 Columbia University Press
All rights reserved

Library of Congress Cataloging-in-Publication Data

Bernal, Victoria.
    Cultivating workers : peasants and capitalism in a Sudanese village / Victoria Bernal.
      p.   cm.
    Includes bibliographical references and index.
    ISBN 0-231-07172-8
    1. Peasantry—Sudan—Case studies.   2. Agricultural laborers—Sudan—Case studies.   3. Capitalism—Sudan—Case studies.   4. Villages—Sudan—Case studies.   5. Agriculture—Sudan—Sociological aspects—Case studies.   I. Title.
HD1538.S73B47   1990
305.5′633′09624—dc20   90-47686
                                                                                                                CIP

Casebound editions of Columbia University Press books are Smyth-sewn and printed on permanent and durable acid-free paper.

Printed in the United States of America
c 10 9 8 7 6 5 4 3 2 1

*To my husband, Tekle,
and to the Sudanese people
from whom I learned so much.*

Contents

Acknowledgments	xi
Author's Note	xv
Maps	
The Sudan	xvi
The central riverain region	xvii

1. Reconceptualizing the Peasantry — 1

   The Fallacy of Household Autonomy — 3
   The Narrow Focus on Agriculture — 5
   The Peasant-Worker Class — 6
   Methods and Research Site — 9
   Overview of Wad al Abbas — 11

2. Class Formation and Agriculture in Sudan — 23

   Markets and Merchants — 25
   Agriculture — 26
   Proletarianization — 33

3. Wad Al Abbas and the Emergence of a Peasant-Worker Class — 41

   Traditional Village Economy — 41
   Mercantile Accumulation and Traditional Agriculture — 53

    Irrigated Agriculture   57
    The Expansion of Off-Farm Work   69

4. Off-Farm Work and Household Economy   80

    Households as Economic Units   82
    The Extent of Off-Farm Work   88
    Off-Farm Work and Household Reproduction   94
    Farming as One Strategy Among Many   106

5. Off-Farm Work and Inequality   112

    Economic Inequality in Wad al Abbas   112
    Mustafa and Amna—A Poor Farming Family   122
    Hassan and Fatma—A Petty Trader's Family   124
    Osman and Katiira—A Family with Working Sons   126
    Abdelsalaam and Hawa—A Wealthy Merchant's Family   128

6. The Labor Market and Household Farm Labor   134

    The Labor Market Hierarchy   135
    Why Do They Farm at All?   142
    Off-Farm Work and Household Farm Labor   146
    The Logic of Household Farming at Wad al Abbas   150

7. Off-Farm Resources, Household Farming Strategies, and Agricultural Productivity   160

    Variation in Agricultural Strategies   161
    Off-Farm Resources and Household Agricultural Production   162
    Farming Strategies and Productivity   171

8. Peasant-Workers and Development   179

    The Peasant-Worker Class and Development Theory   183
    Implications for Agricultural Development Policy   184
    Policy Implications for Wad al Abbas and Sudan   187
    Peasant-Workers and Class Formation   189

Appendix   193

Glossary	203
Measures and Equivalents	207
References Cited	209
Index	219

Acknowledgments

SITTING ALONE at my desk it is easy to feel I did this work all by myself. The truth is many people contributed to this study in various ways. Conducting research in Sudan was not easy. The extreme climate and poor health conditions were a constant test of my physical stamina. The lack of roads and transport, public accommodations for travelers, and internal communications, the shortages of basic commodities, and the sheer precariousness of day to day life in Sudan were at times overwhelming. They surely would have been my undoing were it not for the continual generosity of Sudanese women and men, many of whom did not know me and would never see me again.

My greatest debt is to the villagers of Wad al Abbas. I can only mention some of them here. I especially thank Abdel Bari Abdel Wahid, Risala Mahmoud, the late Omda Al Tayeb and his extended family, the late Shaykh Mohammed Abdulla, Al Khalifa Al Tayeb, Jarra, Bit Moja, Fatma Abdel Maboud, Al Tayeb Al Meki, the late Hussein Al Fadil, Hawa, the Hariri family, Abdel Gadir Al Mamoun, Mohammed Haju, Saiida Al Ballol and her daughter Amal, Nefisa Bit Salman and Mohammed, Zihur Al Mowlana, and Zihur El Amin. I feel very fortunate to have had the chance to live with you a life so different from the one into which I was born.

Translating the varied and complex experiences you shared with me into the discourse of social science was a painful process. Transforming the particular into the general felt like a form of betrayal—relegating you to a nameless, faceless status. In drawing generalizations from your situations and experiences, I have tried to interpret and abstract the important lessons from what you showed me and told me. I hope you will think I listened well and understood.

In Khartoum my heartfelt thanks go to El Fatih, his father Abdullahi Abdelsalaam, his mother Haram, his sisters Fathiya, Asma, Seeham, Nawal, and Iman and his brothers, Usaama and Yasir. In their friendship and concern for me, they went far beyond the bounds of the legendary Sudanese hospitality. I spent my first months in Khartoum living with them and for many months returned to their home whenever I wanted a break from rural life. When I contracted hepatitis and malaria they cared for me. They taught me Sudanese language and social skills that stood me in good stead everywhere I went. To the extent that this study is a good one, it is largely due to the fact that I had good Sudanese teachers and the family of Abdullahi Abdelsalaam is among the most important of them.

Khalid Mubarak Kubeida, El Fatih El Sayyed Humeida, and Marjaana El Sayyed deserve many thanks. Their friendship was extremely valuable to me. They and their extended families in Sennar, especially Amira and Mahmoud, provided me with hospitality and facilitated my research in many ways.

I also thank students and faculty at the University of Khartoum. I am grateful to Taj El Anbia, Chairman of Anthropology, for granting me an affiliation with that department and to the National Archives of Sudan for giving me research clearance. Many thanks to Ali Abdel Gadir, Khalid Hagar, and the students and faculty of the University of Gezira as well.

Friends and colleagues in the United States commented on various versions of this manuscript, offered encouragement, and pleaded with me to stop revising it, at appropriate times. I owe particular gratitude to Ronald Cohen, Henry Rutz, and Dick Franke for their generosity. I am also thankful to Ismail Abdalla, Karen Tranberg-Hansen, Steve Orvis, Lenny Markovitz, Henry Bernstein, Oswald Werner, Jay Stewart, and Errol Balkan.

The National Science Foundation, the Social Science Research Council and the American Council of Learned Societies funded my fieldwork in Sudan from 1980 to 1982. My short field trip in 1988 was funded by the Kirkland Endowment and Hamilton College. I appreciate this support and take full responsibility for the conclusions and opinions in this book.

I am especially grateful to my husband, Tekle Woldemikael for his many contributions to this research. Tekle spent almost two years with me in the Sudan and his own knowledge of the country has been a valuable touchstone for my ideas. He has challenged me intellectually and sustained me emotionally at every stage of the project. Finally, I thank my father, William, my mother, Barbara, and my sister, Lindsay, for their unwavering confidence in me.

Author's Note

THE TRANSLITERATION

In transliterating Arabic words I have tried to reflect colloquial Sudanese pronunciation rather than strictly following classical, literary Arabic. I have left out diacritical markers for simplicity and formed plurals by adding "s" for the same reason.

NAMES

With the exception of historical figures, villagers' personal names are pseudonyms.

The Sudan

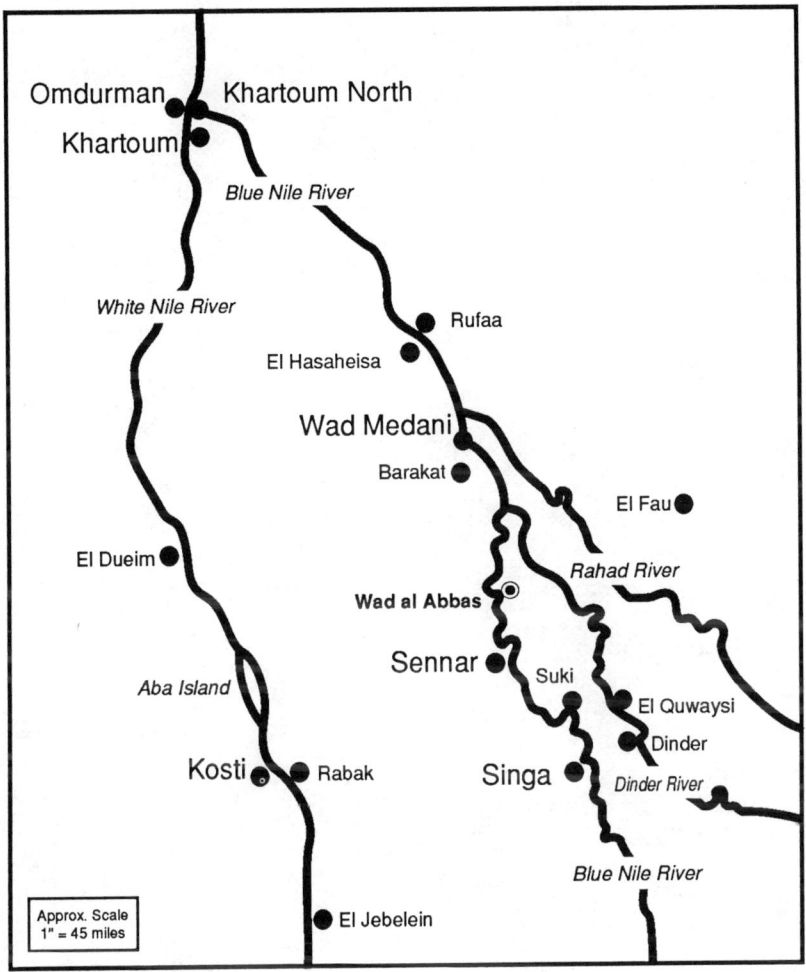

The central riverain region

CULTIVATING WORKERS

ONE

Reconceptualizing the Peasantry

THIS BOOK is about a Sudanese village and about global change. It describes the strategies Sudanese peasants pursue in their struggles to survive, and tells of the growing importance of wage-work in their laboring lives. It is also about the processes of class formation and agricultural development in the third world. I argue that contemporary peasants are an integral part of the working class in many third world societies because this class is reproduced through the combination of peasant production and wage-work. Focusing on agrarian households' links to the wage economy provides new insights into the factors that govern the agricultural behavior of smallholders as well as the process of rural differentiation. By showing that the farming strategies of peasant households are influenced by their participation in regional labor markets, this study challenges current assumptions about peasants, proletarianization, and agricultural development.

The central question of this study is: How does the participation of peasants in wage and informal sector employment alter the dynamics of peasant farming? The answer to this question is explored in a case study of the village of Wad al Abbas in the Blue Nile Province of Sudan.

My interest in the impact of off-farm work on peasant agriculture grew out of two different kinds of experiences I had in Sudan. The most important was the experience of life in rural Sudan and especially in the village of Wad al Abbas. On one of my first visits to the east bank of the Blue Nile, looking for a research site, I stopped at a small village. As usual, I was immediately invited into someone's home. The mud-walled room contained only a few items of furniture, the floors were dirt, there was no electricity, the only running water was a government well at the outskirts of the village. My host drew out a small box from under a cabinet and offered me—fine, imported chocolates—something unavailable even in the capital city! A relative recently returned from working in Saudi Arabia had brought them. At Wad al Abbas I soon learned that villagers were more interested in the exchange rate between the Saudi riyal and the Sudanese pound than in the price of cotton—their cash crop. Although most households own land and produce crops, the majority of villagers can no longer be described simply as farmers.

The other experience that shaped this study was meeting Sudanese and expatriate "experts" concerned with agricultural development in Sudan. Wad al Abbas is not far from the Gezira Scheme, the centerpiece of Sudanese agriculture and reportedly the largest irrigation project under central management in the world. The villagers of Wad al Abbas farm on the Blue Nile Scheme which, like the rest of Sudan's irrigated schemes, is patterned after the Gezira. Sudan depends heavily on cotton from the irrigated schemes as its major export, but cotton yields have been poor and may be declining. Experts come to examine the schemes and make recommendations.

What struck me was that they always referred to the people living on the schemes as "tenants" or "farmers." Yet many of the schemes are located in the most commercially developed region of Sudan and their inhabitants are heavily involved in economic activities besides farming. Researchers tended to view this population mainly in terms of what was being demanded from them—cotton production. They approached agriculture as if it were somehow outside the other economic and social processes changing Sudan; and they treated the schemes in isolation from the regions in which they are located.

From the villagers of Wad al Abbas I learned that agriculture is only one element of a complex and changing local economy. Villagers, then, see farming and their participation in the scheme quite differently from the way scheme managers and some researchers see them. In fact, migrants' remittances, trade profits, subsistence farming, and cotton production all come together in village household economies. Households piece together a living from various sources and individuals try to make the best of difficult circumstances by shifting between farming, petty trade, transport, and wage-work as conditions in their own lives and in the larger economy change.

In my two years in the village I sought to understand the dynamics of farming in the context of villagers' other work and to uncover the relationship between their farming activities and their growing involvement in informal sector and wage employment. In the time since my fieldwork I have become convinced that what I observed was not an isolated case but part of a global pattern. The apparent continuity of rural life and agricultural livelihoods in the third world masks the profound transformations peasants have undergone, transformations which become visible when we focus on the involvement of rural households in wage-work and labor migration, and their connections to the urban proletariat.

The commoditization of inputs to agricultural production and of items of household consumption, along with markets for labor both in and outside of agriculture, alter the basic conditions of peasant production. The role of family labor changes when it can be employed outside the household and when it can be replaced on the farm by hired laborers. Access to capital becomes as important as access to land or to unpaid family labor. Moreover, household agricultural production may become dependent upon resources generated through off-farm employment.

Communities around the world are experiencing processes of declining agricultural autonomy and incomplete proletarianization. But approaches to third world agriculture have not undergone a corresponding transformation. Much research on peasants suffers from a false assumption of household autonomy, viewing smallholders as independent units of production and consumption. It is also common to find agricultural behavior and resources examined outside the context of other economic activities and kinds of wealth. The next section considers these two problems in turn and argues that contemporary third world peasantries are best understood as part of a peasant-worker class.

THE FALLACY OF HOUSEHOLD AUTONOMY

The impact of regional political economy is underestimated in models of peasant behavior that treat the individual farmer or farming household as an autonomous unit. This applies to those who see peasant economy as a non-capitalist mode of production as well as to those who view peasants as penny capitalists maximizing returns to labor and other resources. Both approaches give too much weight to the role of household demographics and farm characteristics, and too little weight to the role of extradomestic relationships and nonagricultural resources, in determining household farming strategies.

The continuing interest in Chayanov's (1966) theories of peasant economy is one example of the persistence of such views (e.g., Millar 1970; Sahlins 1972; Hunt 1979; Donham 1981; Chibnik 1984; Durrenberger 1984). Chayanov saw peasant farming as essentially driven by the internal organization and composition of the peasant household. Moreover, he posited a peasant household with unlimited access to land, depending solely on the labor of its family members. How much of the world's rural population today lives under such conditions? (Indeed Chayanov may have overstated this for Russian peasants in the early twentieth century.)

The farming systems approach (Shaner et al. 1982; Fresco and Poats 1986) is another example of such models. While the farm is viewed holistically, it is analyzed largely as an independent system in its own right rather than as an integral and subordinate part of a larger system.

Such approaches neglect the role of state policies and regional economic structures that determine the conditions under which household resources are managed. As de Janvry has argued: "the social relations that are relevant to an understanding of peasants are external to the family; they are the relations of domination and surplus extraction to the benefit of other social groups" (1981:105).

The larger economic context in which farming households operate is at least as important as factors internal to the household in determining household farming strategies. Household demographics and farm characteristics are meaningful only in relation to the total spectrum of household economy and the position of the household in the regional economic system. A study of New York family farmers concludes that:

> Despite the crucial role of the internal structure of the household, family forms of agricultural production are rooted in the larger economy and society as well. Phenomena external to the household but which affect the structure of household production include factor markets (i.e. for capital, labor, and land inputs), product markets, non-farm labor markets, and public policies. (Buttell and Gillespie 1984:185)

Third world peasant households that purchase agricultural inputs, buy or sell commodities and/or labor must similarly be seen as "rooted in the larger economy."

A further and related problem with some studies of peasant farming is that the analysis focuses too narrowly on agriculture, ignoring peasants' involvement in off-farm work and class relations outside of agriculture.

THE NARROW FOCUS ON AGRICULTURE

Farmers are often analyzed in abstract—removed from the larger economic processes that affect their households and communities, whether or not they as individuals are directly engaged in off-farm employment. The farming systems approach may have a built-in bias toward the agricultural component of household economy (Little 1985). "[There] has been an inclination among anthropologists and other practitioners of 'peasant studies' to agrarianise the countryside, that is, to study and write about rural peoples and their problems as if these involved exclusively agricultural pursuits" (Cook 1984:14).

Viewing peasants solely as farmers, and viewing agriculture as unconnected to the larger economy also has resulted in poor project design. Planners of Sudan's Khashm al Girba irrigation scheme regarded the project, "as an entity in itself and not part of a wider context, whereas the responses of the settlers have emphasized the need to take into account such a wider context" (Hoyle 1977:128–129). Similarly on the Kano River Project in Nigeria: "Non-farming occupations play a crucial role in the economy—a role which has been grossly underestimated in the literature and in the calculations made by researchers and consultants about the income and use of time of Hausa men" (Wallace 1983:282–283).

When they are recognized, off-farm activities are often presumed to be mere adjuncts to farming and, consequently, not subjected to the empirical investigation they warrant. The impact of peasants' off-farm work on agriculture has rarely been addressed beyond the question of labor supply.

Participation in production and exchange relations outside the household can change the conditions and goals of household farming, however. This is overlooked by Hyden, among others, who sees the prevalence of remittances on the part of Tanzanian workers as evidence of the *strength* of "pre-capitalist society" (1980:161). This view ignores the changes in rural economy that might result from dependence on remittances or differential access to them.

The significance of family farm labor and land is modified by the household's other resources, uses of household labor, and sources of income, as well as by the availability and cost of other inputs to agricultural production, such as hired labor, mechanization, improved seeds, pesticides, herbicides, and fertilizers.

From this perspective, it makes little sense to speak of a "peasant economy" with its own independent logic. Rejecting the idea of peasant economy does not mean, however, that we should apply neoclassical economic theory to peasants as if they were capitalist firms. Peasants are neither precapitalist remnants nor penny capitalists. Both these approaches assume too great autonomy for peasants and accord too little attention to the larger system of production under which household production is subsumed. However, these are not the only alternatives.

THE PEASANT WORKER-CLASS

Class analysis offers an improvement over approaches that view peasant farming as a separate mode of production or treat households as autonomous units of production. If a class is defined by its relation to the means of production, this implies a larger context of political and economic relations within which individuals and households operate. Yet class-based approaches to contemporary peasants are problematic, too. A debate has grown up around the conceptualization of peasants, petty commodity producers, peasant-workers, and the proletariat (Bernstein 1988; Roseberry 1976; Gibbon and Neocosmos 1985; Lehmann 1986; Cowen 1981) showing that our classical concepts are not adequate to make sense of real economic forms. In the contemporary third world, the ideal types of the dispossessed proletarian and the independent peasant do not represent the complex reality we observe. For one thing, many of the former peasants under pressure to sell their labor cannot find employment as wage-workers and must instead employ themselves, engaging in petty trade and services in the informal sector. They are thus proletarianized without being proletarians in the strictest sense. In addition, many labor migrants move seasonally between wage-work and their farms. Peasants and wage-workers, moreover, may contribute to a common household economy. Class analyses stressing the distinctions between peasants and proletarians can obscure the links among these economic activities and the people that pursue them.

Proletarians have been defined as the dispossessed, who own no means of production, while peasants, conversely, have been at least partly defined by the fact that they do. It is by expropriating peasants' means of production that capitalists compel them to enter the ranks of the working class. Hence the practice of enclosure in Britain. But in many areas of the third world expropriation has not progressed far. Despite a century of integration into the capitalist system, millions of farmers continue to have access to land on which they produce for themselves as well as for sale or for landlords. This does not mean, however, that these societies are not capitalist or that these farmers are not also subject to capitalist relations of production.

While peasants' access to land has often been seen as a buffer against proletarianization, this is not always true. As capital is accumulated and various means of production are concentrated in the hands of capitalists, and as capitalist production increasingly determines the prices of all goods, the land that peasants possess becomes less and less significant as a means of production. It does not keep them from being dominated by a capitalist class that is effectively able to control the conditions under which peasant production takes place. And it does not preclude the necessity of laboring for wages in order to survive.

This has been particularly clear in southern Africa where possession of land, far from preventing proletarianization, sustains a particular pattern of proletarianization characterized by insecure employment, periodic unemployment, low wages, and circulatory migration (Murray 1981; Parson 1984).

Where subsistence production plays a part in reproducing labor, wages are low for all workers although neither every worker nor every household engages in both wage-work and farming. It is the class, not the individual household, that is reproduced through both subsistence production and wage-work. Furthermore, while subsistence production subsidizes the reproduction of workers employed by capital, wages may contribute to the reproduction of peasants. Subsistence production in many rural areas can no longer be sustained independently of income derived from participation in capitalist relations of production.

The migrant laborer, the petty trader or artisan, and the peasant are engaged in different productive relations and activities. But it is common for individuals to move between these activities over time and for household economies to span them at any given time. They are not discrete modes of production or classes. They are parts of a class of laboring people whose existence and reproduction is predicated on the combination of wage labor/self-employment and agricultural subsistence production—a peasant-worker class.

The peasant-worker class in the third world today emerges from the disruption of autonomous subsistence economies by the expansion of capitalism and the failure of capitalist enterprises to develop sufficiently to employ a large, stable, working class at living wages. Subsistence producers lose the ability to sustain themselves independent of capital without being fully transformed into a proletariat. One of the ways this occurs is through the reduction of their resource base due to the growth of capitalist farms and development projects. Disadvantageous terms of trade between agriculture and other sectors of the economy also undermine the ability of peasants to sustain themselves through laboring on their own resources, as do rising production costs, inflation, and indebtedness. Peasants thus come to depend in part on wage work. Yet wages are too low, and employment too scarce or sporadic to sustain them in the absence of subsistence production. Peasant agricultural production is thus maintained, but it is tied to the wage economy and workers are in turn dependent on subsistence production.

The integration of peasants into the capitalist economy, therefore, is not necessarily a process whereby peasants become landless workers or capitalists. Peasant agriculture persists—but the conditions of production and the behavior of farmers are increasingly determined by the capitalist relations of production in which peasants participate. Peasants' allocation of labor to non-farm work and their access to non-farm income affects their use of agricultural resources. The dynamics of unpaid family labor and production for own consumption are subordinated to the capitalist markets in which peasants are buyers and sellers of commodities, including labor. Peasants' conditions of existence, like those of proletarians, are determined by factors such as wage levels and other conditions of employment, and by the prices of food and basic commodities.

This suggests that the methodological emphasis on farm size and land-based categories, such as small-farmer, large-farmer, and kulak, that has characterized much of the work on third world agriculture is outmoded. The growing participation of rural populations in labor and commodities markets leads to a decline in the importance of land in rural political economy. Household size and composition similarly become less important in agriculture as labor is commoditized. Labor markets, rural-urban migration, wage levels, and patterns of work and accumulation in the informal sector are key factors shaping rural stratification and agriculture.

If present trends continue, the global farming population will be made up more and more of peasant-workers. Policies designed to increase agricultural productivity and rural welfare will likely fail unless they take this into account.

In this book, I demonstrate with empirical evidence the connections between household participation in wage and informal sector employment and farming strategies. I describe the particular conditions in agriculture, commerce, and wage-labor that affect the villagers of Wad al

Abbas and analyze their responses to these conditions. The processes I observed are unique neither to Wad al Abbas nor to Sudan. Because of the many standardized features of Sudan's irrigated schemes, the villagers of Wad al Abbas have a great deal in common with other populations on Sudan's schemes. But like Sudanese inside and outside the irrigated areas, villagers suffer the effects of national economic conditions such as inflation, unemployment, and shortages of food, fuel, and other basic supplies. Moreover, many villagers make their living outside the scheme and thus share working conditions with a broader group of people.

In other parts of the world the changing character of the farming population and the making of the peasant-worker class may be more or less advanced, and vary in its specific effects according to local agricultural conditions, land tenure, household organization, labor markets, industrialization, and features of regional political economy. Peasant-worker communities may vary in terms of whether workers are primarily agricultural laborers or employed in non-farm enterprises, whether they work in their own communities or migrate, and in the relative importance of self-employment in petty commodity production and petty trade. Neither the process of capitalist development nor its effects upon rural producers are homogenous. In fact, the proliferation of diversity may be a key feature of capitalist expansion in the third world, as local histories shape the process of capitalist integration and the characteristics of the peasant-worker class (and other classes) in particular areas. This case study demonstrates that seeing peasants as part of a working class raises fruitful questions about the nature of peasant production, proletarianization, and the course of agricultural development. While answers to these questions must always be local and contingent, the questions themselves have wide applicability.

METHODS AND RESEARCH SITE

Before selecting Wad al Abbas, I traveled in Kordofan and White Nile provinces as well as in the Blue Nile province, considering various possible research sites. This ultimately proved useful in giving me an overview of a large region of Sudan. The village of Wad al Abbas was attractive because it was considered by local people in the area to be a dynamic village, where economic change was occurring rapidly. It is a large village and functions as a small regional center for part of the east bank *(Al Shariq)* of the Blue Nile. Its market is the largest in the area and the village is the site of government offices serving about ten smaller villages. These include a police post, a local court *(Mahkama al Shaabi)* that meets once a week, and a People's Council *(Mejlis al Shaabi)*, the lowest level of local government administration. The administration of a number of local irrigation schemes is also located in the Wad al Abbas office near the village. These institutional links between the village and the state offered forums to be studied.

Wad al Abbas was also an attractive site because culturally and ethnically it is part of the mainstream of Sudanese national culture which historically has been dominated by northern Arabic-speaking groups. Most anthropological studies in Sudan have focused on less powerful, non-Arabic speaking groups. Wad al Abbas also offered me the practical advantage of making Arabic, which I had previously studied, my sole field language. As I spent several months living with a Sudanese family in Khartoum before I selected my site and began fieldwork, I was fairly fluent in Sudanese colloquial by the time I arrived in Wad al Abbas. My language skills continued to improve throughout my two-and-half-year stay in Sudan and I was able to carry out all interviews and surveys without an assistant or interpreter.

Data Collection

I conducted fieldwork at Wad al Abbas from June 1980 to June 1982. The initial part of my study was qualitative; I gathered data through participant observation, open-ended discussions, and formal interviews. In addition to my interviews with villagers, I also interviewed administrators and personnel of the Blue Nile Schemes in the Wad al Abbas and Al Saar offices near the village, and at the main office in Sennar. The qualitative data guided me in designing a household survey to gather data that could be subjected to more systematic analysis and comparison. Due to the large size of the village, a total census was impractical and I decided to survey a random sample of 5 percent of village households. From records of the local government rural council *(Mejlis al Riif)* I compiled a list of household heads (relying on the rolls for subsidized sugar rations, as I expected them to be more inclusive and accurate than lists for taxation or voting purposes). Using the interval method, I obtained a random sample of 47 household heads.[1] Because some of them were polygynous and thus members of more than one household, 53 households were ultimately surveyed.

Each household questionnaire consisted of two parts, one for the male household head and one for the female (usually his wife). While the questionnaires covered a broad ground, questions about employment, property ownership, and agriculture were particularly detailed. Data were gathered on crops planted, area planted, use of hired labor, use of household labor, sharecropping arrangements, cash expenditures, and crop yields and sales for 1980–81 and 1981–82.

I carried out the survey entirely by myself rather than employing research assistants; while this limited the number of households I was able to survey, it gives me great confidence in the quality of the data I collected. Moreover, because I visited each household, most of them more than once, and met many of the households' members, I gained information not specifically included in the questionnaire. I was also able to assess people's answers and probe for more detail on the spot, when necessary.

In January 1988 I revisited Wad al Abbas after five and a half years. This gave me a chance to observe the changes that had taken place in national conditions and in villagers' lives and to discuss them with villagers. I also interviewed head administrators of the Blue Nile Schemes about the status and future of the schemes in general and the Wad al Abbas scheme in particular. The bulk of my analysis is based on data collected between 1980 and 1982, but it is enriched by my awareness of subsequent developments. Wherever relevant I describe the changes I observed in 1988.

OVERVIEW OF WAD AL ABBAS

The village of Wad al Abbas is a community of about 7,000 located in the central riverain region, one of the most developed areas of the Sudan. It was settled largely by Ja'aliyiin and other migrants from the north at the beginning of the nineteenth century. Some of the earliest inhabitants of the village were traders as well as farmers and the village has always been in commercial contact with other regions. Wad al Abbas is not far from the Gezira Scheme where large-scale irrigated agriculture was established by the British in 1925. Irrigated schemes were subsequently set up throughout this region including one at Wad al Abbas in 1954.

The village lies on the east bank of the Blue Nile north of Sennar. A dirt track leads south from Wad al Abbas across the bridge to Sennar, an hour away. Locally owned buses and lorries connect Wad al Abbas and surrounding villages with the town. From Sennar a paved road, completed in 1980, leads to the capital, Khartoum, another five hours away.

The village is bounded on the west by the Blue Nile and stretches along the river. It is flanked on the east by the irrigated fields.

A visitor arriving at the village sees the mud houses, cultivated fields, and wandering cattle, the men in their white robes and turbans, and the women shrouded in their flowing *towb*s (body veils). The stillness of the village in the heat of the day and the slow movements of the people suggest the tranquility and security of unchanging traditions and the steady rhythms of rural life. Nothing could be further from the truth. The isolated appearance of the village belies its integration into the world economy. During my two years of residence in the village, I was to discover again and again that, while removed from the centers of power, the village is in no way removed from outside economic and political forces. Wad al Abbas is linked to the national economy of Sudan through the consumption of goods from the market, the marketing of its labor and products, and through the state-run Blue Nile irrigation scheme on which villagers produce cotton, Sudan's primary export. The fates of villagers are ever more dependent upon national and international economic conditions.

Yet the lives of villagers also depend upon local institutions of marriage, household, and family, which remain important lines of access to resources and control over labor. Although villagers are involved in capitalist relations of production and exchange, most households continue to produce a subsistence crop, and unpaid family labor contributes significantly to agriculture as well as to household chores.

Ethnicity

If one can speak of a northern Sudanese "national" culture with local variants, the inhabitants of Wad al Abbas are part of it. Their basic social norms and values of generosity, religious piety, family obligation, sex segregation, and their marriage and residence practices conform to a general pattern I observed in Khartoum, Gezira, and White Nile provinces. Village ceremonial life—weddings, circumcisions, naming *(simaya)*, and funerals *(bika)*—also fits within this general pattern. Other accounts of northern Sudanese life are generally consistent with my own observations (Ibrahim 1979; Salih 1985; Boddy 1989). Variation between villages or between village and town results as much from the different consumption levels of various economic groups as from cultural differences.

Most of the inhabitants of Wad al Abbas are Ja'aliyiin, descendants of the founders of the village. According to historians the Ja'aliyiin are Arabized Nubians (Hassan 1973; Holt and Daly 1979); Ja'aliyiin themselves, however, emphasize their Arab descent and Muslim cultural heritage. In its most inclusive sense the term Ja'aliyiin refers to all the Arabized sedentary riverain populations of northern Sudan (Hassan 1973; Holt and Daly 1979). Northerners have historically dominated Sudan's political economy and exerted cultural influence over other regions. The Ja'aliyiin are, in this sense, the quintessential Sudanese in relation to whom others are defined as ethnic minorities. Ja'ali (s.) identity is associated with power and prestige in the Sudan today. Ja'aliyiin hold many government offices and civil service positions and figure prominently among the national elite. For this reason people of mixed origins are likely to emphasize their Ja'ali ancestry. This is certainly the case at Wad al Abbas. In addition to Ja'aliyiin, Wad al Abbas is made up of descendants of local nomadic peoples such as Kawahla and Sherifa.

The village is composed of neighborhoods *(fariiqs)* usually named after one man and roughly corresponding to his descendants, including males and females, and their in-married spouses. Although people are still aware of their different descent, there are no ethnic divisions to speak of within the village, except for the case of a few descendants of slaves with whom other villagers do not intermarry. Villagers commonly refer to themselves as one family *(ahal)*, though they are aware that they are not descendants of one ancestor. Through long co-residence and intermarriage they have coalesced into one community with common customs and way of life. With very few exceptions, residence at Wad al Abbas is based on birth or marriage; thus, villagers' sense of being quasi-kin of one another is grounded in reality.

On one edge of the village there is a separate settlement of Nuba from western Sudan who are not regarded as part of the village. Many of them came as laborers to dig irrigation canals when irrigated schemes were established in the area in the 1950s. They remain culturally distinct from the villagers and it is possible that some are non-Muslims. As non-natives of the area they have no rights to local land; they work as agricultural laborers and sharecroppers for villagers, produce and sell illegal alcoholic drinks, and make shoes and other items for sale.

Religion

All residents of Wad al Abbas proper are Muslims. Their religion and its association with Arab culture are important sources of identity and pride for them. This both reflects and contributes to their increasing integration into the Sudanese nation and the Arab world. Like the great majority of Sudanese Muslims, the people of Wad al Abbas are Sunni Muslims. Villagers are affiliated with a variety of *tariiqa*s (religious orders), the most popular being the *Khatmiya* (also called *Mirghaniya*), the *Gadiriya*, the *Semmaniya*, the *Tayibiin*, and the *Yagoubab*. Religion is a frequent topic of daily conversation among old and young, men and women, educated and uneducated, wealthy and poor. Nonetheless, many villagers, even today, are Muslims (accepting Mohammed as the prophet and believing in one God) but know little of the Islamic faith. For example, I encountered women who did not know the words to their prayers, they simply performed the motions. Religious practices and beliefs continue to evolve as villagers increasingly adopt a more orthodox, literate, cosmopolitan understanding of Islam. In this they are part of the broader movement of Islamic fundamentalism in Sudan and abroad.

Religious changes at Wad al Abbas are linked to its changing position in the regional economy. Villagers say that it was the urban, educated employees of the irrigation scheme who first exposed them to many aspects of orthodox Islam and revealed to them the "backwardness" of their own way of life. For example, prior to the establishment of the scheme, alcohol was produced and consumed publicly in the village, something unthinkable today. Labor migration and trade in urban areas and, more recently, employment in Saudi Arabia have contributed to a growing orthodoxy among villagers. Religious education in government schools also gives Wad al Abbas youth a different religious training than that traditionally provided by village *fakis* (holy men).

As the stricter, more devout, and comprehensive understanding of Islam spreads in the village, certain practices that existed within or alongside Islam are coming to be defined as "against Islam" and are abandoned or suppressed. For example, the custom of young men whipping each other at weddings had disappeared by 1980, and a movement is now under way to eliminate wailing at funerals and to limit funeral gatherings to three days. Weddings and funerals are occasions of great social and ritual significance at Wad al Abbas; such changes are thus further indications of the profound transformations taking place in village life.

Economy

By many criteria Wad al Abbas is a developing village. The village has a small health station and a small veterinary station. Schools were built beginning in the late 1950s and now include boys and girls elementary and intermediate schools. There are plans for a higher secondary school. One villager had attained a Ph.D. by 1980, and several villagers are currently studying abroad as well as at the University of Khartoum. Wad al Abbas has produced some teachers and civil servants, too. In 1981, three young village women graduated from high school—the first women of Wad al Abbas to achieve this.[2] These three women were also the first women to hold salaried jobs. They all became teachers in the village schools.

In 1974, twenty years after the irrigation scheme was established, a pump system to supply water to villagers was put in place. Before that time, villagers depended on the Blue Nile for their water as they still do when the pump is broken or out of fuel. In the early 1980s the pump only ran briefly twice a day for villagers to collect water from faucets in their own or neighboring courtyards, but by 1988 villagers had running water most of the day. Local efforts to bring electricity from the plant at the Sennar dam up to the village were finally successful in 1988. In anticipation of the event people had installed lights and those who could afford to had bought other items, including televisions, fans, and electric irons. All of these projects including the schools involved village self-help *(awn al zaati)*, generally in the form of cash donations, as well as government support.

Compared to neighboring villages and to its own past, there is evidence of rising prosperity at Wad al Abbas in terms of consumer goods such as radios, cassette players, and even a number of televisions, as well as the many purchased items in daily household use such as china and pots and pans. A few of the wealthiest villagers own cars and trucks. From straw and mud huts *(gotiyas)*, some villagers have converted to mud-brick houses and others to fired brick. There are even a few cement houses with iron grillwork comparable to urban dwellings. One and two generations ago people subsisted on a diet of the staple food, *dura* (sorghum) made into flat bread *(kisra)* and porridge *(lugma* or *asiida)* with vegetable sauces. Nowadays many villagers enjoy a more varied diet including meat and fruit. These changes in housing and diet, however, also reflect a shift from household consumption based on household production to an economy largely based on purchased goods.

The livelihood and prosperity of villagers are dependent on economic opportunities outside the village and even outside the country as men migrate to work in Saudi Arabia and other places. Wad al Abbas is becoming ever more of a residential area, a place where women raise children and young people receive some basic education before leaving for higher education or employment, and a base that men can retreat to in economically inactive periods such as holidays, sickness, old age, and unemployment. The most economically active elements—young men—are quite often physically absent from the village though economically they are a vital part of it.

Standards of living in the village clearly are improving in respect to housing, diet, and access to public services. Indeed between 1982 and 1988, brick houses, appliances, televisions, and furniture became considerably more common. Yet this prosperity seems largely due to an intensification of labor on the part of villagers who experience ever-increasing pressures to maintain or improve on the gains they have made. Time has become money. As one village youth commented, "In the past if you went to a man's house at dusk you would find him at home. Now, they are all out—running after money."

The cycle of daily life and community social and ritual activities are more and more geared to the exigencies of labor migration and rigid work schedules. Funerals and weddings are truncated compared to past practices. And the most recent signs of women entering full-time wage-work (although their number is too small to speak of a trend) suggests that households are reaching further into their labor pools in an effort to increase their cash incomes.

The differences of wealth among villagers are growing; a few live in comfort while many live in poverty. These differences are based in the relative positions of villagers within the emerging class structure of Sudan, not on any internal economic structure or patronage system within the village itself. One villager's wealth is not necessarily derived through monopolizing local resources or exploiting fellow villagers. Rather, villagers compete in the national arena for control over productive resources. At present only a small minority of villagers, professionals and merchants, have been successful enough to become part of the national bourgeoisie. The majority of villagers are part of the diverse peasant-worker class, managing by their wits to eke out a living in the worst of times, and to acquire some new possessions in better times, but never accumulating capital.

Inequalities in wealth among villagers are not highlighted by formal social distinctions, however. As neighbors, kin, and members of the village community, people are linked in informal relationships as social equals even though they are very aware of differences in their economic circumstances. Wealthy villagers do not set themselves apart culturally.

Among the younger generation, those in their twenties and early thirties, however, social and economic differentiation is more pronounced. This is largely due to the recentness of class formation, so that there is a much broader range in the education, occupations, and social contacts outside the village among the younger generation. The elder generation are mainly either traders or farmers and, while divided by differences of wealth, their social worlds overlap to a great degree.

Social and Political Organization

The community has few positions of authority beyond the general principle of a father's authority over his sons and the authority of the eldest brother over the others. Women are under the authority of fathers and brothers and, after marriage, husbands. Women past child-bearing age gain a great deal of autonomy and a degree of authority over younger relatives. Outside of immediate kin relationships, power generally means influence rather than command over others.

The kinship classification system used by villagers is of the "Sudanese" type where distinct terms designate almost every relative. Nanda's (1984:239) comments apply to Wad al Abbas—"Although most groups using this system tend to be patrilineal, there is also evidence of ambilineality. This distinguishes these systems from other patrilineal systems." Genealogies and ethnic descent are traced patrilineally, but relatives on ego's mother's side are recognized by kin terms that mirror those on the father's side. At Wad al Abbas it is generally assumed that people feel emotionally closer to their mother's side of the family while maintaining more respectful relations with their father's side of the family.

There is little emphasis on tracing genealogies back beyond three generations. Genealogy is valued not so much for relationships to dead ancestors as for determining kinship to the living. Within three generations even youngsters are quite knowledgeable about their precise kin ties to their relatives. However, in conversation kin ties are used loosely and there is a tendency to collapse them, making relationships closer. Thus *akhuy* (my brother) often refers to a cousin and *wad ammi* (my cousin) may refer to the offspring of a parent's cousin or another more distant relative, while *khaali* (my maternal uncle) could mean any man related through the mother's side. There are no patrilineages (*khasm bayts*) at Wad al Abbas, nor any corporate groups beyond the level of the household and the extended family.

There is an egalitarian strain running through the society so that among women and among the men of one age group, there is little hierarchy. Thus, although women are under the authority of men, and younger men under the authority of older men, in practice, due to sex segregation and age peer-grouping among men, in many social contexts there is no authority structure. Diffuse social pressure seems to substitute for this in maintaining order and conformity.

There is, however, a group of elder, respected men, *nass al kubaar* (old or big people), who are looked to as mediators in disputes and whose opinions and advice are sought and likely to be heeded. They include the *imam* of the village mosque, who is a descendent of the village founder, al Faki Mohammed Ali Wad al Abbas, and some of the wealthier villagers, though wealth alone does not bring this kind of respect. Moreover, while wealth and religious knowledge are sources of prestige, they cannot be directly translated into authority over others, only to influence. Visiting patterns reflect social status; those of lesser prestige pay their respects by visiting social superiors. On Fridays and holidays men tend to gather at the *diwaan* (guest house) of the eldest man of their *fariiq*.

Those positions or institutions that carry with them legitimate authority were introduced into the village from outside by the agricultural scheme and by colonial and post-colonial governments. As a legacy of the British colonial Native Administration, Wad al Abbas still has a *Shaykh* who collects government taxes and is in some sense a local liason between government and villagers although the *Mejlis al Shaabi* now has taken over much of this role. The *Shaykh* is invested with little real power by the government and is unsalaried. The position is inherited and the degree of informal power associated with it appears to depend on the leadership capabilities of the individual and the wealth and prestige of his family. The *Shaykh* is also the president of the local court *(Mahkama al Shaabi)* in which capacity he has authority in court cases and for which he receives a government stipend.

Sex segregation is an important principle of social life at Wad al Abbas. It is achieved through the seclusion of women and through separate social spaces for men and women. Young girls up to the age of ten or so have considerable freedom of movement. After that and until the end of child-bearing women are restricted from public places and contact with unrelated men. Old women, in contrast, move freely and even travel outside the village on their own. Seclusion is particularly pronounced when a girl reaches marriageable age and becomes most strict once she is engaged. As a consequence, engagement often results in girls being withdrawn from school. In the early months of marriage, a bride remains secluded at home and must not go out even for important obligations such as funerals. After the first year of marriage, however, a wife begins to have more freedom of movement than an unmarried girl and this increases as she matures.

There is a strong association of women with private space and men with public. The market, the fields, and anywhere outside the village are basically the realm of men while women's domain is generally limited to the confines of the village and within the *hosh* (courtyard) which demarcates domestic space. While men are focused out of the village, through economic, social, and political ties—women's world is the village. Many women have rarely if ever been to Sennar and then usually only to the hospital there. In contrast, many men commute to Sennar on a daily basis for trading, and many have traveled widely within the Sudan. Women travel occasionally to visit their children living elsewhere, attend weddings and funerals, or to make the *haj* (the Islamic pilgrimage to Mecca). Significantly, by 1988 a number of village women had made trips abroad, accompanying or joining their husbands employed overseas. One woman had herself been employed along with her husband in Yemen. Thus, gradually, women are being drawn directly into the world beyond the village. However, for the most part it still reaches women indirectly through the accounts of male relatives and the effects of male labor migration and other economic changes on their daily lives in the village.

Within the village, men lounge in front of the neighborhood shops, crowd the market and the courtroom, and loiter chatting in the paths. Women of child-bearing age do not themselves buy or sell in the market or shops and they do their chatting behind *hosh* walls. It is more acceptable for women to go out at night as the darkness in a sense makes the public private and they are not exposed to the view of men. Women do not pause long when they meet in the path but walk purposefully toward their destination. They generally use smaller, hidden paths while men take the main thoroughfares through the village. Except for the very old, women do not pray at the village mosque but in their own *hosh*. And even the older women do not enter the mosque, they pray in the courtyard outside it.

Public meetings are attended exclusively by men. Women do not attend the court and, if necessary, the judges meet them at home. Though by law 25 percent (6 members) of the Local People's Council must be women, in practice women members do not attend its meetings. Sudanese women have voting rights and some village women vote in government elections. When a villager ran for a regional office in 1981 women formed part of his entourage, to reach women voters in other villages.

Sex segregation is least evident within the immediate family where husbands and wives, and even more so, brothers and sisters, have intimate relaxed relationships. This proved important during fieldwork as it allowed me to interact informally with men in many of the homes I visited as well as to observe male-female relations. In this and other ways I was treated differently than an unrelated Sudanese woman would have been. However, the close ties of kinship and marriage that link so many villagers to each other allow men and women considerable familiarity with one another in informal social settings.

Even within the family, the sexual division of labor is pronounced, however, and men and women do not usually cooperate in tasks but rather perform complementary activities. If guests are present, sex segregation becomes nearly complete—the women in one area of the *hosh*, the men in the *diwaan* (guest house). This is always the practice at large gatherings such as weddings, funerals, sacrifices *(karaama)*, and other ceremonial occasions. Not all households can afford to have a *diwaan* or even an enclosed courtyard *(hosh)*. Most make some effort to achieve seclusion for women by erecting screens of thornbrush, burlap, or sticks. For some poor households, the *hosh* is more a concept than a reality, and the women, although in private space, are in public view.

The women of Wad al Abbas like other northern Sudanese women wear the *towb*—a large (6 meter) piece of cloth covering head and hair but not the face and falling to the ankles. Quite often it is of sheer colorful cloth. The *towb* is worn over a short sleeveless dress or smock. Older women wear heavier black or white *towb*s and often leave their breasts bare underneath, wearing only a kind of baggy shorts *(sirwaal)*. While townswomen only put on the *towb* when going out or if a male guest is present, Wad al Abbas women keep their *towb*s at least partially on (as it is not fastened but only wrapped, it is always slipping), or, at the very least, right next to them at all times. Small girls just wear dresses and girls from about eight to fourteen wear a short shawl *(terha)* covering head and shoulders down to the waist.[3] (In towns, girls do not switch from the *terha* to the *towb* until they are eighteen. This may be related to the later age of marriage in town since girls of marriageable age should be secluded as much as possible.)

In towns, many men have switched to Western dress for public activities, wearing the traditional *jellabiya* (robe) and *aimma* (turban) informally at home. In Wad al Abbas, however, men as well as women wear traditional dress. Salaried employees and university students, who wear Western dress when out of the village, wear *jellabiya*s in Wad al Abbas. Farmers, artisans, and other poor people wear shorter robes *(araagi)* most of the time, donning a *jellabiya* only for Friday prayers or special occasions. Friday prayers are an important communal ritual in Wad al Abbas as most men pray together in the village mosque, not at home or in their neighborhoods as they do at other times.

While people are dispersed through work, education and, less commonly, marriage which take them out of the village, they maintain their social membership in the community in various ways, sometimes going to great lengths to do so. One communal obligation is funerals. Every adult villager is expected to offer condolences to the family of the deceased. People return to the village from far away to attend funerals. In 1982 some men spent £S70 (about $78) to come by taxi all the way from Khartoum, pay their respects at a funeral, and get back to their jobs. Those unable to attend funerals make the rounds of all the houses who have experienced losses immediately upon their return to the village. The entire village celebrates the major holidays, *aeed Ramadan, aeed al Duhiya,* and *al Mowlid* together through such practices as communal prayer in the open field near the mosque, a communally attended *zikr* (ritual chant and dance) and a meal of rice and milk. For three days during *aeed Ramadan* and *aeed al Duhiya* the village paths are crowded as everyone, even youngsters, seeks to greet and congratulate every household in the village. At such times the village teems with life as its farflung members come home for the holidays.

Villagers have a sense of Wad al Abbas as a community and a society. They talk about *"nass al barra"* (people of the outside), such as teachers who are not from the village, or city people. One woman on learning that I paid rent to the family in whose *hosh* I lived, exclaimed "Do *we* rent? Have we become like *nass al barra?"*

Villagers are aware of great changes in their lives and in their place in the world. While sometimes they are nostalgic for the life of the past, other times they emphasize what they see as positive changes. One of the oldest living villagers in 1982, Haj Hussein, who said he was 92 at the time, reflected this way: "We are very happy. Things have become good. We have become like this [putting his hands together with fingers interlaced] in the world. After this, is separation possible?"

NOTES

1. Taking every 20th household head, I selected an initial sample of 54 out of the total of 1080 household heads. Seven of the 54 were eventually eliminated—

three no longer had households in the village, one was never available, and three were not truly members of Wad al Abbas proper, two of these worked in Wad al Abbas but did not have households there and the third was from a community of Nuba associated with the village.

2. There is also a woman whose mother is from Wad al Abbas but who was raised and educated elsewhere by her father. She was recently the first Wad al Abbas woman to get a university degree. During fieldwork she stayed with her mother in the village whenever the university was not in session. She eventually married a Wad al Abbas man and accompanied him to Saudi Arabia.

3. I found the *towb* impractical and difficult to manage. Therefore, I adopted the *terha*. This satisfied villagers' standards of propriety. But it had the disadvantage of emphasizing my young age and signifying a girlish, child-like status. On the other hand, my somewhat novel dress and appearance indicated that, although I was properly modest, I was not behaving exactly as, and therefore should not be treated exactly as, a Wad al Abbas woman. The nature of my research required that I go places and do things that a Wad al Abbas woman should not do. The mixed message of my attire was thus appropriate for my ambiguous social status, indicating only partial conformity to the norms of seclusion.

TWO

Class Formation and Agriculture in Sudan

CULTURALLY, THE Sudan is often described as "Afro-Arab," a blending of identities and peoples. The name Sudan itself comes from the Arabic word for "black." Sudan has been seen as a microcosm of the African continent with Arabic-speaking Muslims in the north and Christians and animists of various African linguistic groups in the south. While the Sudan has always been part of Africa and her links to Arabia predate Islam (Hassan 1973), the West has exerted influence over Sudan since at least the nineteenth century. Arabic is the official national language and serves as a lingua franca throughout the country. English is used in southern government departments and some other official contexts, and remains the language of instruction at some universities.

Sudan was incorporated into the Ottoman empire under Turko-Egyptian rule *(Al Turkiya)* (1820–1885) and had a brief period of independence under the Islamic state of the Mahdi (1885–1898). Sudan's unique experience of Anglo-Egyptian colonial domination (1899–1955) contributed to the spread of Islam and Arab identification in the north, even as it connected Sudan to the Western world and spread Christianity in the south. Unlike many of its neighbors, the Sudan never had a substantial European settler population, however.

The Sudan is tied to the West, especially the United States, through aid, loans, technology, and other key imports. At the same time, its historical links to Arabia remain vital and have taken on new importance as oil revenues have made Saudi Arabia and the Gulf states important sites of capitalist expansion in the region and sources of aid and foreign investment for Sudan.

The Sudan is the largest nation in Africa, covering close to one million square miles. It is also one of the poorest, least developed nations in the world. In the early eighties the Sudan's record on satisfying basic needs was "close to the lowest in the world" (ILO 1986:10). Recent figures put the per capita income of Sudan's 22.6 million people at about $320 per year (World Bank 1988). Life expectancy is forty-nine years. Fewer than half the nation's school-age children were enrolled in primary school in 1983–84 (ILO 1986). The country is a net importer of foodstuffs as well as all finished goods (Ali 1988). The economy remains heavily based in agriculture which accounts for 35 percent of its gross domestic product (World Bank 1988) and 80 percent of its exports (ILO 1986). Sudanese agriculture can be broken into three main sectors—smallholder rainfed farming, mechanized capitalist rainfed farming, and irrigated agriculture. Up to now most of Sudan's agricultural development efforts have focused on the irrigated sector which produces the country's major export crop, cotton. The population is predominantly rural, and the only city with more than 500,000 inhabitants is the capital, Khartoum. In 1980, 8 percent of the labor force were employed in industry, while 71 percent were employed in agriculture and 21 percent in services (World Bank 1988).

In the past decade the Sudan has suffered drought, flood, famine, civil war, and military coups. The country has been host to hundreds of thousands of refugees from the neighboring nations of Ethiopia, Zaire, Uganda, and Chad, and has experienced large internal population movements due to famine and the civil war between the north and the south. Sudan's foreign debt and trade deficit have grown exponentially. By conservative estimates inflation averaged 33 percent a year between 1980 and 1986 (World Bank 1988). Per capita food production declined between 1979 and 1987 (Africa Recovery 1989). Shortages of basic commodities such as bread, flour, sugar, and fuel are common. In January 1989 UNICEF estimated that 90 percent of the rural population lacked access to safe water (Africa Recovery 1989).

These statistics measure the symptoms of Sudan's poverty and underdevelopment. To understand their causes we must look at the historical processes that shaped the Sudanese economy, giving rise to these adverse conditions and to the peasant-workers who struggle to survive them.

A general narrative of Sudanese history is still being created through systematic studies of different social and economic formations. What follows is not a summary of established consensus but an interpretation based to some degree on ambiguous and conflicting sources. Emphasis is on northern Sudan which dominates national political economy, and particularly on the central riverain region, which is the setting of our case study.

MARKETS AND MERCHANTS

The Sudanese economy has historically been characterized by both a strong subsistence base and a strong indigenous commercial class linked to the international market system. Saeed argues that:

> prolonged occupation and continued transfer of the social surplus either to the seat of power of the Islamic Empire via Egypt between the 7th and 15th centuries; or later to the seat of the Turkish empire through its regional power center in Egypt; meant the gradual and consistent incorporation of the central Nile valley and its hinterlands into the dominant world system of appropriation in its mercantilist capitalist phase and later in the competitive capitalist phase. (1982:83)

One of the early centers of trade was the Funj (1504–1821) capital of Sennar, located in what is now the Blue Nile Province, where an international commercial community was operating by 1700 (Spaulding 1984:40). O'Fahey and Spaulding (1974) date the rise of a Sudanese bourgeois class to the middle of the eighteenth century at Sennar, associated with the slave trade. In the early nineteenth century, Sennar, with an estimated population of 10–15,000, may have been the largest town in the Sudan (Issawi 1966:464). (It declined rapidly after that, however, and the present town of Sennar was later established nearby.)

International (and to a lesser extent domestic) trade was the primary source of state revenues for the Turkish and Mahdist governments of Sudan in the nineteenth century (Ahmed 1977:31). From the second half of the nineteenth century northern Sudanese, especially the Ja'aliyiin, but also the Danaqla and the Shaiqiya, dominated trade in the Sudan (O'Fahey and Spaulding 1974:102). "At first the most flourishing branch [of commerce] was the slave trade, developed by Europeans as a byproduct of the ivory trade but soon taken over by northern Sudanese . . . [B]y the mid-1860s all Europeans had withdrawn leaving the field open to Arab settlers and slave raiders and traders" (Issawi 1966:469, 486). A considerable number of Sudanese capitalists today trace their family wealth back to the slave trade (Mahmoud 1984).

Slave trading required no reorganization or control over production by merchants, and it has been argued that the other primary export goods from Sudan through the nineteenth century, ivory, ostrich feathers, and gum arabic, were not so much commodities cultivated or manufactured for exchange, as natural resources (Sellin 1980:622; Bjørkelo 1984). While trade developed prior to British colonial domination, it had little impact on traditional forms of production and did not bring about the commoditization of means of production and consumption (Sellin 1980:621; Abdelkarim 1988). At the turn of the century "the greatest part of the economy . . . was restricted to production for subsistence alone" (Niblock 1987:11).

AGRICULTURE

The role of slave labor in Sudanese agriculture is a subject of debate. It is not clear how widespread slave ownership ever was or to what extent slave labor permitted free Sudanese landowners to withdraw from cultivation. Slave labor apparently enabled farmers to cultivate larger areas of land than they otherwise could have. Many of the Sudanese slave raiders and owners, however, were nomadic pastoralists (McLoughlin 1962) who had little interest in establishing agricultural estates. While the history of slave-holding is sometimes invoked to explain the use of hired laborers and sharecroppers by Gezira farmers in recent times, no evidence is presented to back these claims (Culwick 1955; Warburg 1978). O'Brien notes that "there is little evidence to support the widespread contention that slave labor contributed significantly to surplus production in Sudanese agriculture" (1980:48). It is likely that slave labor merely supplemented family labor, since agricultural productivity probably was too low to support large numbers of non-producers (O'Brien 1980:513). In nineteenth century Shendi, for example, the use of slave labor "did not mean that cultivation was not a household undertaking. Only those better off did not work themselves and they also kept their women folk out of the fields. However slaves enabled their owners to work less on the farm, and if they so wanted, to devote more time to other activities, such as trading" (Bjørkelo 1989:66). Little is known about the social or economic dynamics of slavery within Sudanese households. McLoughlin (1962) states that slaves shared the same living standard as their masters, though he does not reveal the basis for this conclusion. While slaves were used in agricultural production, it seems the greatest impact of slavery was the mercantile wealth some Sudanese derived from the slave trade, which was largely export-oriented.

Despite the emergence of an indigenous merchant class, no class of landlords developed in Sudan until as late as the mid-twentieth century. Arable land remained abundant and merchants accumulated trade capital rather than land. From early times the state played a significant role in creating private estates, however. It was the practice of the Funj Sultans to make land grants to holy men (O'Fahey and Spaulding 1974). These religious leaders generally did not act as landlords but settled on the land with followers who offered political allegiance and paid tribute (Adam 1977:34). Through tribute and slave labor religious leaders accumulated wealth and expanded their political power with the support of the Funj Sultans.

After defeating the Funj, Turkish authorities continued the practice of granting private estates to individuals. Unlike the religious leaders, these lay grant-holders did not settle on their lands but remained in Sennar, operating as absentee landlords to tenants with whom they had no social or religious bonds (Awad 1971). Even so, as late as 1924 the British Foreign Secretary could state that practically the whole of the riverain land (land near the Niles) remained in the ownership of those who worked it (in Awad 1971:224). Outside the riverain areas, land remained even more plentiful.

Rather than acquiring large estates, merchants primarily exercised indirect control over agriculture through credit relations which allowed them to appropriate the products of peasant farmers without owning the land. The penetration of capital via mercantile capital thus had consequences for the course of agricultural development. Unlike the industrial capitalist, the merchant does not operate under constant pressure to develop the forces of production and in fact transfers surplus from production to circulation (Kay 1975:95).

After the industrial revolution in Europe, merchant capital in the periphery became "the agent of industrial capital" (Kay 1975:100).

> The expansion of the home market, the export of agricultural commodities and goods from the national capitalist sector, and the distribution of imported commodities, all provide a basis for a rapid increase in the circulation of commodities, and for an economic strengthening of the merchant class . . . This expansion of the trading function provides the basis both for ensuring that non-capitalist production is increasingly directed towards the needs of capitalist reproduction, and for the growth of a *trading class dependent upon the maintenance of the restricted and uneven*

> *development that imperialist penetration has produced.* (Taylor 1979:226; emphasis in original)

Such was the case in the Sudan until at least the middle of this century.

The expansion of Sudanese agricultural production was stimulated by the extraction of tribute, taxation, and the availability of consumer goods through the market. But agricultural productivity remained limited by labor supply and the organization of production by peasant households. Farmers lacked the wealth to command significant amounts of labor, slave or hired. Those who had wealth, the merchants, generally did not invest it in agricultural production but in expanding and diversifying their commercial activities.

One consequence of this was that to pay taxes, purchase consumer goods, and meet their other needs Sudanese peasants became increasingly dependent on merchants for credit. This dependence was reflected in the *shayl* crop-mortgage system which became widespread in the Sudan and is still practiced in some areas although it is now illegal (Barnett 1977; Saeed 1982). In a standard *shayl* agreement, the farmer receives grain or other household supplies and perhaps cash from the merchant during cultivation. In return the farmer pays the merchant back in kind at harvest. While the farmer borrows at a time when grain is scarce and prices high, he repays at harvest when grain is abundant and the price has dropped. Since *shayl* agreements require the farmer to return an amount in kind equivalent to the *value* of goods borrowed at the pre-harvest price, he therefore must return more than he borrowed.[1] This can result in effective interest rates of 200 percent or more (Shaw 1966a). *Shayl* has been described as entailing:

> an intricate social relationship between the moneylender and the cultivator . . . He is in constant touch with the farmer and is a built-in feature of village society. He is on hand to receive and grant requests for loans on short notice, for productive and social purposes, on the apparently flimsy basis of the security of a farmer's crops . . . there develops a kind of moral obligation on the part of the lender to help out in time of need . . . (Shaw 1966a:D58)

While Sudanese farmers were slowly drawn into petty commodity production, the organization of agricultural production remained the same: the peasant household farming its own land. Duffield (1978:6) writes of a village on the Blue Nile upstream from Sennar: "although a cash crop economy had existed in Maiurno since the beginning of the 1920s it was dominated by a form of production amongst the peasantry based on relations of personal dependence within the household." Nonetheless, the entry of merchant capital as a component of household economy linked farmers in relations of dependence that extended beyond their families to the merchant class.

Through the early decades of this century peasant production expanded and religious and tribal leaders extracted tribute while merchant capital acted as an intermediary between the peasants and international capital (Khalafalla 1981a). Trade, not agriculture, was the primary basis of accumulation by the indigenous bourgeoisie. Under the Anglo-Egyptian Condominium (1899–1955) trade increased as infrastructure and communications improved, and Khartoum and lesser commercial and administrative centers grew (Niblock 1987:19–20). By 1912 railroads linked central Kordofan and Khartoum with the Egyptian border to the north and the Red Sea coast to the east.

It was the British colonial government and foreign capital rather than the Sudanese merchant class that profoundly reorganized agricultural production. Starting around 1910, the British Condominium government granted land concessions to foreign companies such as the Sudan Plantations Syndicate. Irrigated schemes, most notably the Gezira Scheme, were developed through British-Syndicate cooperation. The establishment of the Gezira Scheme in 1925 marked the beginning of a new era in Sudanese agriculture.

Awad (1971) argues that an incipient feudal class was prevented from developing in the Sudan by British land policies which restricted the rights of estate owners and limited their share in tenants' crops. According to Awad (1971:221), the incorporation of large private estates into the Gezira Scheme with only minimal compensation was "the final blow to the landed aristocracy of the Sudan." The degree to which the Sudan ever had a "landed aristocracy" is open to question. Awad (1971:220) himself notes that "the Turkish administrators and most of the foreigners who came to settle in the country found it easier to get rich by trading in ivory or slaves or through corruption than by investing in land." He gives no reason why this would not be equally true of Sudanese.

While by some reports, prior to 1925, land in the Gezira was concentrated in few hands (Barnett 1977:90), others (Issawi 1966:497; O'Brien 1980) hold that the Gezira was populated mainly by small cultivators working their own land. Land registers apparently show large holdings partly because different Gezira families sometimes registered their lands collectively to save on registrations costs (Miskin 1950). Moreover, some local leaders registered communal lands as their own property at the establishment of the scheme (Niblock 1987:15). There was evidently no problem of land shortage or fractionalization in the Gezira or the Blue Nile Province at the beginning of this century (Awad 1971:223). Only along the Nile north of Khartoum where cultivation requires irrigation was land pressure a problem. However, there was considerable migration from this region to land-abundant areas. Moreover, inheritance practices were inclusive rather than exclusive, so that while people had smaller and smaller shares in land, they were not disenfranchised (Bjørkelo 1989; Omer 1985). These practices, in turn, made land sales virtually impossible because numerous co-owners were involved (Miskin 1950), thus serving to restrain the process of commodification of land.

In any case, to whatever extent a class of large landholders had developed prior to British rule, colonial policies restricting land transfers and limiting rents made "the ownership of large estates of negligible social and economic importance" (Awad 1971:221).

From the start, the agricultural development of Sudan was determined by external interests. Pressure from the Lancashire Cotton Growing Association in Britain led to the development of the Gezira as a cotton-growing region while the irrigated growing season was defined by the Nile water needs of Egypt (Gaitskell 1959). The colonial government encouraged the expansion of irrigated cotton production by offering land licenses to wealthy Sudanese willing to invest in pumps. Most of the licensees were not large landowners but merchants who relied on loans to establish their schemes and invested their profits in non-agricultural ventures (Khalafalla 1981a, 1981b). As on the Gezira, cultivation on the pump schemes was carried out by peasants who were the nominal owners of individual irrigated tenancies (*hawashas*). The organization of the schemes and the relationship between licensees and cultivators were set out in government regulations.

In the early 1950s, world cotton prices shot up, precipitating a "cotton-rush" as private mercantile capital leapt into irrigated cotton production (Osman 1958). No other investment promised as quick and large returns as the cotton schemes at that time (Shaw 1966b). Between 1944 and 1957 the number of pump schemes grew from 372 to 2,229, most of them in the Blue Nile Province which at that time included the Gezira (Osman and Suleiman 1969). The boom was short-lived, however, as world cotton prices fell at the end of the 1950s. The state, which had fostered and regulated the schemes, took on a greater role as private capital moved to more lucrative enterprises. The Gezira Scheme was brought completely under government management in 1950. In 1956 the Sudan gained its independence, and control passed to the new national government. When private financiers became reluctant to back cotton production, the Sudanese government created the Agricultural Bank to finance the schemes. Throughout the 1960s cotton prices remained low and, to assure their continued operation, the government began to nationalize schemes as the licenses of private owners expired. The compensation these owners received allowed them to invest in other areas of the economy, and many shifted to mechanized rainfed agricultural schemes (discussed below).

While the British promoted cotton production to supply their textile mills, post-colonial governments have relied on cotton as the primary source of foreign exchange, and have continued to expand the irrigated area following the same general pattern. Today, the irrigated area in Sudan totals 1.7 million hectares, more than the total irrigated area in the rest of sub-Saharan Africa combined (Horowitz 1989). An estimated two million Sudanese live or work on the schemes. (For comparative materials on Sudan's schemes see Barnett 1977; Sorbo 1985; O'Brien 1984; Salem-Murdock 1989.)

Tenants on the irrigated schemes have been alternatively viewed as comparatively privileged (Tait 1978; Voll 1980) and as an exploited group (Barnett 1977; Ali 1983; Said 1968; Collins 1976). The main advantages scheme residents have generally enjoyed, relative to the rest of the rural Sudanese population, are access to public services such as water, health care, and education, as well as being favorably situated in relation to markets and transportation. On the other hand, they are constrained to produce within an agricultural system over which they have little control and from which they cannot extricate their land. Farmers on the schemes are required to cultivate cotton regardless of the returns and have no control over cotton sales since the crop passes directly into the possession of scheme management at the harvest, with farmers receiving a share of the sales proceeds. "The tenancy has become in effect a permanent heritable leasehold, for which the government abstracts, by way of its cotton monopsony, a share rent" (Robertson 1987:95).

Much has been made of tenants' use of hired labor and sharecroppers (Warburg 1978; Brausch 1964; Nigam 1977) as an indication of their privileged position. However, the labor required on the irrigated tenancies is greater than most farm families can themselves provide (Bernal 1990; Khalafalla 1981b). Moreover, incomes and food production from a tenancy have generally been deemed inadequate for the subsistence of a family (Sørbø 1977; Barnett 1977). While wealthy merchants who own tenancies can afford to hire workers to cultivate for them, this is not true of most tenants. In the mid-1960s it was estimated that only 1.7 percent of tenants on the Gezira Scheme (who are generally better off than tenants on other schemes) relied primarily on hired rather than family labor to work their land and functioned as managers rather than cultivators (Niblock 1987:87). A 1969 study found that absentee owners accounted for half the tenants in one area (Taha 1973); however, such a high figure is likely due to overlooking the involvement of family members besides the nominal owner in cultivation. Moreover, while many tenants employ some labor, they do so under conditions imposed by the state which ultimately profits from the organization of labor inputs by tenants. Thus O'Neill (1978:10–11) argues: "Contrary to what an analysis of legal relations would suggest, the principal contradiction is not between tenant farmers on the one hand and migrant labourers on the other, but between all these producers of surplus-value and the small class of wealthy tenants and commercial capitalists who utilize the neo-colonial state to exploit them." Historically, tenants have allied themselves with workers, and tenants took part in the Sudanese labor movement of the 1950s and '60s (Khalafalla 1981b).

In addition to the tenant population, large numbers of seasonal agricultural workers were needed in the irrigated areas, especially for the cotton harvest. From the 1930s, the colonial state took active measures to stimulate if not coerce migration from other regions, especially Kordofan, to the schemes (O'Brien 1988:140).

Like large-scale irrigated agriculture, mechanized agriculture was first begun directly by the colonial government and expanded under successive post-colonial regimes which promoted private investment in this area. As Saeed notes, "The central role of the state in promoting capitalist development in 20th century Sudan is the most important aspect of colonial and contemporary political economy" (1988:187). By 1966 mechanized schemes covered 1,297,000 *feddans* (544,740 ha) (Khalafalla 1981b). Mechanized rainfed agriculture is another way in which the merchant class entered production under favorable conditions created by the state. Merchants are granted government leases at nominal rents to large tracts of rainland which they cultivate mainly by seasonal wage-labor. These licensees generally have little interest in the land, cultivating it until fertility declines and moving on to another tract or into another enterprise (O'Brien 1981). In recent years, mechanized agriculture has been the most rapidly expanding branch of Sudanese agriculture.

Outside the schemes, agriculture remains predominantly the unmechanized rainfed farming of smallholders. They supply much of the seasonal agricultural wage-labor to the irrigated and mechanized schemes.

PROLETARIANIZATION

Agriculture

The emergence of a working class in Sudan was not an immediate consequence of its integration into the world market or the rise of an indigenous bourgeoisie. Wage-labor existed alongside slave labor in Sudan at the turn of the century (Warburg 1978). In a curious twist, some slaves were allowed by their masters to undertake wage-work and even labor migration with the provision that they remit a portion of their wages (McLoughlin 1962). As early as 1908 the cash-paid labor force in the Three Towns (Khartoum, Khartoum North, and Omdurman) numbered 15,000 (Niblock 1987:21). But the smallness of the wage-labor market is reflected in the fact that when the earliest irrigation scheme was established at Zeidab in 1905, wage-laborers were imported (some all the way from America) to work on it (Gaitskell 1959:50). The attempt to operate the scheme with wage-labor was quickly abandoned. Even by the late 1920s, the early years of the Gezira Scheme's operation, there was virtually no wage-labor market for it to draw on (O'Brien 1980:16) and the scheme was structured around peasant farmers who would cultivate their holdings with family labor.

From the early 1900s, the colonial state had begun to pressure Sudanese to sell their labor, through expanding the cash economy by promoting consumer goods, levying taxes in cash from rural producers, and paying local leaders cash stipends (Niblock 1987). Initial efforts were particularly aimed at converting slaves and ex-slaves, who were often landless, into wage-laborers (McLoughlin 1962). Workers were needed for infrastructure projects such as railroads and dams, and for canal construction and harvesting on the irrigated schemes where unpaid family labor had proved inadequate to meet the peak demands of cotton production. Sudanese were also needed to work as police and at the lower levels of administration. However, it was not until the period between 1925 and 1950 that a wage-labor force began to emerge out of the Sudanese peasantry (Khalafalla 1981a:71). Even so, it was almost exclusively composed of seasonal agricultural laborers who were peasants and nomads themselves.

The emergence of an agricultural labor market thus did not entail the creation of a class of landless workers, nor did it herald the rise of a class of capitalist farmers. Gezira tenants were the primary agricultural employers. In some areas, such as Gedaref, seasonal wage-labor was used in small-scale commercial cultivation before the mechanized schemes (Abdelkarim 1988), but family labor remained the backbone of Sudanese agriculture. The majority of farmers lacked the means to hire much labor. Until the 1950s wage-labor was used in Maiurno village only on a few large farms of *shaykhs* and merchants who had formerly owned slaves (Duffield 1978). Elsewhere the use of hired labor by village farmers remained limited at least through the 1960s (Khalafalla 1981a).

The formation of a class of free wage-laborers dispossessed of their means of production had not progressed far by mid-century. Peasants and nomads retained control over resources, and much wage-work, like that on the agricultural schemes, was based on the seasonal labor of workers who returned to their own fields or herds in the off-season. Wages in many cases were below subsistence levels since they often paid seasonal labor and included benefits in kind as well as cash.

The rapid expansion of irrigated schemes in the fifties and mechanized capitalist agriculture in the sixties greatly increased the demand for seasonal wage-labor. The expansion of these schemes did not deprive rural communities entirely of land, but reduced their resource base, making non-agricultural income, often from wage-labor on the schemes, a key to their survival (Elhassan 1988:163–164). In addition to claiming the resources of peasants and pastoralists for the state and its merchant licensees, mechanized schemes, which generally produce sorghum and sesame, compete with the products of smallholders in the market, thus undermining petty commodity production. The organization of agricultural commodity production by the state and, under its auspices, by merchants, thus pressured peasants to supplement their subsistence production with income from agricultural wage-labor rather than through expanding commodity production themselves (Niblock 1987; O'Neill and O'Brien 1988:11). British policies such as outlawing cotton cultivation in some areas outside the irrigated schemes also fostered this.

> The existence of a substantial body of poorer peasants—impelled to seek temporary labour on the agricultural schemes at harvest-time, but able to support themselves elsewhere for the rest of the year—was a necessity for the Sudanese economy under the Condominium. (Niblock 1987:82–83)

These conditions led to the emergence of a peasant-worker class as ever larger numbers of people could be reproduced only through combining family-based subsistence production with wage-labor or self-employment in petty commercial activities. Clearly to the extent that the subsistence economy remains viable outside the circuit of capital, wages may merely supplement subsistence production. But, external pressures such as taxes, inflation, indebtedness, ecological degradation, and the encroachment of various schemes, have made wages and/or cash incomes from petty trade and services increasingly essential to the survival of rural populations in Sudan. Subsistence production is thus sustained by wage-work, and subsistence-supported labor in turn subsidizes capitalist production.

Even by independence in 1956, a substantial number of Sudanese farmers, perhaps the majority, were unable to subsist solely on their own farms and engaged in seasonal wage-labor to survive (Niblock 1987:82). Their numbers grew in the post-colonial period.

> With the increase in the population, progressive desert encroachment and the growth of mechanized farming, the rural population was pressured into migrating from their communities to urban and semi-urban areas. According to the 1973 census, the scale of migration from one province to another within the Sudan trebled between 1956 and 1973 (Mahmoud 1984:28)

By the mid-1970s conditions faced by the rural masses limited their ability to meet their subsistence needs through their own production, compelling them to depend on the market for income and consumption (O'Neill and O'Brien 1988:13). The ranks of seasonal wage-laborers in scheme agriculture alone had swelled to nearly one million by 1980 (Khalafalla 1981b). Moreover, it is predicted that, "absolute numbers of agricultural labourers will continue to increase for a long time . . . because the nonagricultural sectors are, and will likely remain, relatively small" (Mohamed 1986:97).

The Sudanese peasantry has not differentiated into landowners and a landless proletariat; landowners work for wages and wage-workers own land. For example, 78 percent of a sample of settled agricultural laborers in the Gezira had access to land as tenants or sharecroppers in 1983, while 92.5 percent of a sample of seasonal laborers there had access to land elsewhere (Abdelkarim 1988:150–51). Furthermore, farmers on and off the schemes are often sellers as well as purchasers of labor (Elhassan 1988).

The growing dependence of rural Sudanese on non-farm income is related to low incomes and productivity in agriculture, while the prevalence of labor migration is due in part to the uneven development between regions, and between urban and rural areas. This unevenness itself is largely the result of colonial and post-colonial government policies that concentrated investment in infrastructure and services in a few areas, leading private investment to be concentrated there as well.

While the ranks of seasonal agricultural laborers continue to grow and labor migration has become a way of life for Sudanese, opportunities for permanent employment have not kept pace.

Industry

The industrial development of Sudan was extremely limited under British rule, and handicrafts actually declined as a result of competition from imports. It may be that "the colonial system had a vested interest in keeping the country as a supplier of agricultural commodities and as a market for its own manufactured goods" (Mahmoud 1984:52). Certainly, the colonial government followed a policy of agricultural rather than industrial development. Local private investors had limited capital and, since domestic industry had little protection from imports, it was not an attractive investment (Mahmoud 1984). Until 1944 there were only ten industrial establishments in the nation and manufacturing accounted for less than 1 percent of the GDP at independence in 1956 (Mahmoud 1984). The manufacturing sector was small and composed of small-scale units of production, employing only .03 percent of the working population in 1956 (Niblock 1987:45). Aside from the agricultural schemes, capital primarily dominated commerce and finance rather than production (Khalafalla 1981b).

Construction, both government and private, was an important area of wage-employment throughout the colonial period. The colonial government was a major employer in administration, services, and transportation as well. In the early decades of this century, however, there were so few urban wage-laborers that foreign workers (mainly Egyptians, Greeks, Armenians, and West Africans) as well as prison labor had to be relied upon (Al-Shazali 1988a). But as rural Sudanese increasingly were drawn into the cash economy, some of them, particularly those from Western and Central Sudan where there was little agricultural wage employment, began to migrate to urban centers for work. The proximity of the Gezira Scheme to Khartoum allowed workers to shift between urban and agricultural wage-labor depending on employment conditions. Because wage-work was limited, and urban food supplies unreliable, it was also common for urban workers to maintain farms on the outskirts of Khartoum, leaving their wage jobs for farming during the rainy season (McLoughlin 1970). Overall, the urban labor supply was linked to agricultural conditions, swelling when rains or harvests were poor, shrinking when harvests were bountiful.

The demand for wage-labor fluctuated with world market conditions, declining as a result of the depression, rising as Sudan was used to supply goods and manpower for Britain's war effort in the 1940s (McLoughlin 1979). Many Sudanese soldiers joined the ranks of workers when they were demobilized after World War II (Niblock 1987). But the urban labor market was too small to absorb them, and the colonial government encouraged soldiers and rural migrants to return to their homes or migrate to other agricultural areas (Al-Shazali 1988a).

Among government workers, the Sudan railway workers are most notable. They numbered 20,000 in 1946 and were concentrated in Atbara (Khalafalla 1981b). They, rather than manufacturing workers, formed the nucleus of Sudan's labor movement. It is significant that from the start their demands were not limited to their own working conditions but aimed at national economic and political reform. Moreover, the trade union movement sought links with the rural population, particularly tenants on the irrigated schemes whom they helped organize (Khalafalla 1981b).

In 1956, urban workers numbered 346,253, but this figure includes self-employed artisans and domestics as well as workers in services and manufacturing (Niblock 1987:90–91). Moreover, domestics, whose remuneration often consists mainly of room and board, constitute the largest single occupational category in this population. Salaried workers numbered additional 52,854 that year. An estimated 90 percent of the Sudan's population remained engaged in some form of peasant agriculture or pastoralism (Niblock 1987:46).

Commerce rather than production remained the primary avenue of accumulation. Transport was also a field of investment and was closely associated with trade, as merchants invested in freight vehicles (Niblock 1987:42; Mahmoud 1984). Between 1945 and 1956 the number of registered trucks and lorries increased from 2,718 to 10,798, about two-thirds of which were privately owned (Niblock 1987:42).

At independence, the indigenous merchant class was in a position to be well represented in government and administration. From these offices they expanded their economic activities within the existing social and economic framework (Niblock 1987). The state continued to play a key role in the economy, investing in irrigated and mechanized schemes, establishing a number of factories, and expanding the infrastructure. Public and private industry grew, but was concentrated in Khartoum.

This pattern of development has persisted. In the early 1970s, 77 percent of large-scale industry and 73 percent of the industrial labor force were located in the capital (Al-Shazali 1988b:242). Commerce, transport, mechanized schemes, and real estate continue to be the main areas of private investment and accumulation. Even by 1980 only 8 percent of Sudan's labor force was employed in industry (World Bank 1988) and from the colonial period the state has remained the largest single employer in the country. Only about one in three migrants to urban areas finds employment in the productive sectors (agriculture, industry, transport, and electricity), the rest end up working in trade and services (Mohamed 1986). The informal sector absorbs much of this labor because people create their own employment, working as petty traders, errand boys, and odd-jobbers.[2]

There has been considerable continuity in the Sudanese economy in terms of the key role of trade as the basis of accumulation and class formation, and in terms of the large numbers of Sudanese who remain engaged in family-based subsistence production. In the 1980s, "the overwhelming majority of Sudanese—about 80 per cent—are farmers, farm workers or nomadic herdsmen" (O'Brien and O'Neill 1988:19).

For the mass of the population conditions grew increasingly difficult over the 1970s and 1980s. Per capita income fell and inequities in the distribution of income increased (ILO 1986). Through the 1970s the real wages of seasonal agricultural laborers declined (O'Neill 1978). In that decade the government embarked on major investments in infrastructure and development projects based on foreign investment and loans, incurring immense debts that posed a burden on the economy. Sudan's debt approached $9 billion by 1985, growing to nearly $14 billion by 1989. Inflation rose to an effective rate of 60 percent per year in the early 1980s (Niblock 1987:283), and between 1980 and 1984 the cost of living index for "low salaried workers" tripled (ILO 1986:12). Sudan also has suffered the devastation and financial burden of two civil wars between the north and the south since independence. The government reportedly spends a million dollars a day in the current conflict.

Since the 1970s the Gulf states have become an important source of foreign investment in Sudan, primarily in finance and agricultural schemes. At the same time, development in the Gulf has attracted Sudanese labor as jobs overseas offer an alternative to the low incomes and unemployment in Sudan. International labor migration to the Gulf began among professionals and skilled workers who were soon followed by less educated migrants from rural areas. Through the 1970s large numbers of Sudanese sought employment abroad, with the greatest number going to Saudi Arabia. Between 1968 and 1975 the number of Sudanese working abroad rose from 900 to 52,000 and was estimated to be 100,000 in 1980 (Birks and Sinclair 1980). As approximately 78 percent of these migrants leave Sudan illegally their true number is not known (Galal-al-Din 1988). Estimates for 1982 put Sudanese emigrant workers at one million or more (ILO 1984). This migration is almost always temporary, and migrants generally leave their families behind.

Without industrial development in Sudan, the number of wage-workers outside of agriculture, where work is temporary and seasonal, remains relatively small. Those who have been identified as working in Sudan's formal sector (ILO 1976), where employment is presumed to be more stable, are a minority of workers. Significantly, even these workers are often employed on a temporary or casual basis.

> Casual and temporary employment in Sudan is not . . . confined to the construction industry; or even to those other large establishments and government departments in which the pace and intensity of production fluctuate with the season. Even in large-scale manufacturing establishments—where employers are expected to be most concerned with developing "durable" relation [sic] with workers—"non-permanent" employment is extensive. (Al-Shazali 1988b:241)

In 1973, 60 percent of the unskilled workers employed by a group of public industrial concerns in Khartoum were temporary workers, not counting the casual workers employed strictly on a daily basis (Al-Shazali 1988b:241). Out of an estimated 33,000 employed in manufacturing in Khartoum, 31,000 (94%) were in temporary jobs (ILO 1976:365 in Al-Shazali 1988b).[3]

The instability and insecurity of employment, even within the public, formal sector, shows that many urban workers, like their rural counterparts, cannot rely solely on wage-work to sustain them and their families. Even more telling are the low incomes in wage-work and the decrease in real wages between 1970 and the 1980s. While the minimum wage rose from £S13.90 per month in 1970 to £S35.83 in November 1983, an increase of 158 percent, inflation between 1970 and 1984 is estimated at 700 percent (Al-Shazali 1988b:256).

Conditions in agriculture make wage-work or self-employment in the informal sector vital to Sudanese peasants, while labor market conditions encourage workers to maintain links to their rural communities and to keep an interest in rural land and family farming. "It is only in relatively exceptional circumstances when an unskilled migrant is lucky enough to find himself in a well remunerated position that he tends to give up interest in the productive activities he used to pursue in his home village" (Al-Shazali 1988b:259).

Sudan's working population, urban and rural, constitutes a peasant-worker class whose members are reproduced through family-based subsistence production, wage-work, and small-scale self-employment. While capital accumulation and development in Sudan will eventuate in a full-fledged proletariat severed from its peasant roots, there is little to suggest that the peasant-worker class is merely a transitory phenomenon. The conditions that gave rise to this class will likely persist for some time. In the near future, the ranks of peasant-workers will grow as more isolated or resilient communities are drawn into the wage economy.

The case study of Wad al Abbas sheds light on the nature of the peasant-worker class and the significance of this class for agricultural development. The study also reveals how the harsh and uncertain macroeconomic and political conditions described here affect the lives of rural people and the strategies they use in their quest for survival and prosperity.

NOTES

1. This arrangement allows merchants effectively to obtain interest on their loans without breaking the Islamic prohibition against charging or paying interest.
2. Unfortunately there is comparatively little documentation of the history and development of Sudan's informal sector.
3. Legally employment for more than 90 days requires a permanent position. (It is not clear from these studies what the rate of turnover is or the degree to which temporary employees are working more than 90 days.) In any case, workers in temporary positions have little protection under labor laws and can be legally fired at will.

THREE

Wad al Abbas and the Emergence of a Peasant-Worker Class

TODAY THE people of Wad al Abbas travel hundreds of miles for work, their fields offer them little sustenance and they forsake the comforts of family and community to make a living in unfamiliar places. But it was not always like this. This chapter tells how the relationship between the village and the region has changed over time and how village economy has been restructured in response to pressures from outside the community. The first part of the chapter provides an overview of the history of Wad al Abbas from its founding in 1808 to the 1950s. The second part focuses on the rapid economic changes the village has undergone since 1954 when an irrigated scheme was established there. It describes the present agricultural system and the growing dependence of villagers on off-farm work.

TRADITIONAL VILLAGE ECONOMY

This section is primarily an oral history based on the statements of villagers themselves, often without the benefit of other documentation. The limited published sources on the area are used to supplement the information gained from villagers wherever possible. Villagers' accounts of traditional agriculture are probably most accurate for the period immediately preceding the establishment of the scheme. However, descriptions of traditional riverain and rainland agriculture in the Gezira and further north present much the same general pattern, indicating great continuity over time and space.

The village of Wad al Abbas was established about 1808 by the Faki Mohammed Ali Wad al Abbas who received the land as alms from the Funj regent, Mohammed Adlan. In the title deed, the land is described as "cultivable land . . . being waste rainland" (Holt 1969:4). Apparently, the steady cultivation of the land dates from this time. The area had probably been used by semi-nomads and nomads on a seasonal basis before that time.

Villagers say the region originally was inhabited by a people called the Anaj. According to O'Fahey and Spaulding (1974), the Anaj were the people of Soba who lived around the confluence of the two Niles and the northern Gezira. The Abdallab, who conquered them, and pushed south in the sixteenth century, came to be referred to as Anaj by some (O'Fahey and Spaulding 1974:23–39). These Abdallab/Anaj ultimately dominated the Butana (the plain between the Atbara river on the east and the main Nile and the Blue Nile to the west). It is not clear which of the two Anaj peoples inhabited the area of Wad al Abbas. According to some villagers, there were also Mahas (a Nubian ethnic group) farming in the region when Mohammed Ali Wad al Abbas settled there. Some Mahas became followers of the *faki* and joined the village. Mohammed Ali himself, born in Wiheib in northern Sudan, was of Ja'ali (and Abbasi) origin. His *nisba* (patrilineal genealogy), which traces his ancestry back to Arabia, is kept in the village. Ja'aliyiin and Nubians joined him, as part of a larger population movement from north to south, motivated by the abundance of riverain land and arable rainland upstream. The main attractions of Wad al Abbas appear to have been good farm land and the water supply offered by the Blue Nile. The *baraka* (holy power, literally "blessing") of the *faki* may also have attracted some of the settlers. A few local pastoralists of Kahli and Sherifi origin joined the village early on, as did others from the Gezira. During the upheavals of the Mahdiya (1881–1898), a few more Ja'aliyiin and some Danaqla joined Wad al Abbas.

Agriculture

The basic characteristics of the agricultural system at Wad al Abbas were: 1) a diversity of land types; 2) a diversity of crops; 3) concentration on the production of food crops for direct household consumption; 4) availability of new land and no landless; 5) unequal distribution of land holdings; 6) low level of technology; 7) the household as the productive unit; 8) the use of slave, and later hired, labor primarily as a supplement to family labor and largely restricted to wealthier villagers; 9) credit available to farmers through the *shayl* system of crop-mortgage and many farmers dependent on this credit; and 10) a three-to-four-month slack period between March and July, when all crops had been harvested and new ones could not yet be planted, during which villagers engaged in other activities, such as long-distance trade.

Land Tenure. The Funj land charter does not specify the measure of the land granted to Mohammed Ali Wad al Abbas; its borders are defined in relation to the holdings of others. Although the charter grants Mohammed Ali and his heirs sole ownership of the land, the *faki* allowed his followers to clear land and own it in their own right. Individual land rights established through the clearing and cultivation of land were widespread in northern Sudan (Bolton 1948:188). Once he cleared it, a man farmed the same land year after year. Eventually it was inherited by his heirs. Successive generations farmed the land they inherited and cleared new land as they needed it.

Islamic inheritance practices, whereby daughters can inherit property in shares equivalent to half those of sons, and widows inherit an eighth share of their husband's wealth, were probably followed at Wad al Abbas from early on, as they are today.[1] In this way, rather than through clearing and cultivation, women could come to own land and pass on rights in land. In practice today, such land is usually managed for a woman by her brothers, husband, or sons. However, it seems that women were more actively involved in agricultural production in the past and therefore exercised land ownership in their own right more than they do today (Bernal 1988a).

A farmer's holdings consisted of three types of land, cultivated in season: rainland *(bilad)*, riverbank floodland *(jeref)*, and river island land *(jazira)* (the last being land on islands that emerged from the river as the water level dropped). This pattern of riverbank and *jazira* cultivation was also practiced in the Northern Province (Ibrahim 1979:35). Many of the early settlers of the village had thus been farming this way even before coming to Wad al Abbas.

In addition to farming for themselves, villagers cultivated the fields of the *faki* and the *khalwa* (Islamic school) teachers. This practice was widespread and apparently has not totally died out. Describing the cultivation of Funj land grants to religious leaders, Awad writes: "As the custom is today, the land was cultivated by the students and followers of the shaykh working on a part-time basis, and the crops which they produced were the chief source of sustenance for the people of the *masid* (religious school)" (1971:219).

Soon after the founding of Wad al Abbas, the Funj empire declined and Sudan came under Turko-Egyptian rule. Under this regime, Abdel Rahman Ab Ziig of Wad al Abbas was made *nazir* (regional native authority) of a large area and given a land grant of 20 *jeda'a* (46.2 ha). He was a trader as well as a farmer, and one of the wealthiest men in the village. Individuals and local leaders who cooperated with the Turks, "were assimilated into the administration of the newly established provinces and they succeeded as [sic] establishing themselves as middlemen, buying the peasants' products cheap and selling them to the markets of the enlarged towns" (Adam 1977:36). This was probably the case at Wad al Abbas. It is difficult, however, to understand the importance of the land grant to Nazir Abdel Rahman, when land at Wad al Abbas was apparently abundant and individual land rights were recognized by the community.

There were no landless at Wad al Abbas. However, *jeref* and *jazira* land were more limited than rainland so that, as the village grew, the only access to such land was through inheritance, and eventually there were households without river lands. Land sales were recorded in commercial centers of the riverain region (Spaulding 1982), but land was not bought or sold at Wad al Abbas before the irrigation scheme. As one man put it, "Every man had land—not for money, you cut the trees and farmed, not for a *girish* (cent) or a *ta'rifa* (half cent)."

Agricultural Schedule and Crops. Wad al Abbas is located roughly at 14 degrees north latitude and 34 degrees east longitude on an arid savanna plain. The local soil is high in clay content like the soil of the Gezira. Rainfall in this climatic zone can reach 25–30 inches (635–762 mm), but the village lies close to the border of the semi-desert steppe zone to the north, where annual rainfall is below 10 inches (254 mm) (Best and De Blij 1977:508). According to villagers, annual rainfall has been declining, at least over the past few decades. Today vegetation is sparse and the area is virtually treeless, though villagers say there used to be many trees which they used for fuel and cleared off land for cultivation. All of the trees were eventually cut down to establish the irrigated scheme.

Villagers' accounts and published sources vary somewhat as to the yearly agricultural schedule and the range of crops cultivated. This may be due in part to regional climatic differences—rainfall starts earlier and is heavier to the south, so rainfed agriculture starts earlier at Wad al

Abbas than in some areas of the Gezira. Yearly fluctuations in the timing and amount of rainfall also affect agricultural schedules and the choice of crops. Furthermore, new crops, like tomatoes and cotton, were introduced to different areas at different times.

The year is broken up into three seasons: *khariif* (the rainy season) from July to October, *shita* (winter) from November to February, and *saif* (summer) from March through June. Villagers timed their sowing in the rainland with the onset of the rains. The *jeref* and *jazira* land were planted later when the river receded. Rainland was irrigated by *teras*, ridging fields to retain rainwater. Brausch et al. (1964:19) describe this practice in the northern Gezira. There was never a *saagia* (ox-driven water-wheel) at Wad al Abbas though there were several in the vicinity—at Fariiq Hamdun, Gundal, and Wad al Jamaali. During fieldwork a *saagia* still in operation was observed not far away at Al Deim al Musheikha on the west bank of the Blue Nile. No plows or draft animals were used in cultivation. Farming was carried out manually with hoes (*jiraya*) and digging sticks (*selluka*). Hill gives a good description of these agricultural practices near Sennar (1970:30–31).

The agricultural year began in July when the main subsistence grain *dura* (Sorghum vulgare) was planted on the rainland. There are many different varieties of *dura* and several were grown at Wad al Abbas. In addition to growing *dura* for grain, villagers cultivated a type called *ankoliib*, prized for its sweet stems which are chewed like sugar cane.

Farming continued through March or April when the last of the *jazira* crops were harvested. In October, *dura* was harvested from the rainland; then as the river subsided, the *jeref* and, after it, the *jazira* were planted. These produced the same variety of crops: watermelon, maize, wheat, beans, lentils, cucumber, squash and gourds, tomatoes, fennel, and pidgeon pea (Cajanus cajan). Most likely such staples as okra, jew's mallow (Corchorus olitorius), and purslane (Portulaca oleracea) were also grown. Describing *jeref* farming at El Gedid in the Gezira, Randell (1958:27) reports that different crops were cultivated at different levels as the river subsided. These food crops were produced for household consumption, although surpluses were sometimes bartered or sold.

Some villagers apparently had begun to cultivate commercial crops as well, before the irrigated scheme required all farmers to do so. A few villagers mentioned planting sesame in June or July on the rainland and harvesting it in August or September, and others spoke of cultivating cotton on the rainland and on the *jazira*.

Sorghum was by far the most important crop grown at Wad al Abbas, however. It was the basis of the subsistence diet and the main product sold by villagers to meet their cash needs. It was also the basis of the *shayl* credit system.

Grain Supply and Productivity Levels. Productivity levels in traditional agriculture at Wad al Abbas are difficult to assess. It is doubtful that farmers consistently produced large surpluses. Villagers remembered food shortages in the past, and a famine on the east bank of the Blue Nile due to insufficient rains is recorded in 1949 (Gaitskell 1959:262). But many villagers also remembered selling grain. They said traders from Omdurman, Khartoum, El Kamliin, and Wad Medani came regularly to Wad al Abbas to purchase sorghum. The village is located in what was Sudan's grain belt before much of the land was put into irrigated cotton production (Jefferson 1949).

Some families had holdings of 50 *feddans* (21 ha) or more which were said to yield 100 *ardeb* of sorghum, an average yield of 2 *ardeb/feddan* (660 liters/hectare).[2] Writing in the late 1950s, Randell (1958:37) reports that rainfed sorghum yields averaged 2 *ardeb/feddan* at El Gedid in a good year, but he notes that such a good year was rare at that time and he suspects that the area had experienced declining productivity due to decreasing rainfall. Before irrigation, the semi-nomads farming on the Gezira had good crops in 2 out of every 5 years (Beer 1955). Thus, while yields might have averaged 2 *ardeb/feddan* at Wad al Abbas in a good year, when taken over longer periods, average yields were probably lower.

The riverain areas reportedly produced grain surpluses under the Funj (O'Brien 1980:30) and under Turko-Egyptian rule. The Turks, and later the British, exacted tribute and taxes from Wad al Abbas in grain. Grain supply at Wad al Abbas fluctuated but it would seem that, by and large, the village was more than self-sufficient.

At Wad al Abbas, surplus grain was more often stored than sold. Storage was one way of coping with variable yields, and villagers preserved grain by burying it in underground pits *(matmuras)*. Grain could apparently be successfully stored in *matmuras* for eighteen months or more (Darling 1951). The *Faki* Mohammed Ali Wad al Abbas stored large amounts of grain to feed the hungry who might come to him for aid in a bad year, and villagers also relied on each other for such help. Even now, unless forced by pressing cash needs, most villagers prefer not to sell surplus grain but store it in their houses or storehouses *(makhzins)* until after the next harvest. Today, however, most households fail to produce sufficient grain for their own consumption and are purchasers rather than sellers of sorghum.

Households traditionally cultivated a variety of crops and were largely self-sufficient in food production. They sold grain to cover cash expenses such as the purchase of oil, sugar, coffee, tea, shoes, and clothing. In addition to cultivating food, villagers gathered edible plants. Today, women gather several different types of green vegetables, including a wild okra and a bitter green leaf they call *mollayta*. In the past, there may have been more varieties available for gathering.

Organization of Agricultural Production. The basic unit of production was the household. A man farmed with the help of his unmarried sons. Even after marriage, sons might continue for some time to farm family holdings with their father. But ultimately, perhaps when his own sons were old enough to provide labor, each son would establish his own farm.

There is controversy over the extent of women's participation in farming. Some villagers maintain that Wad al Abbas women never farmed and that they have always considered such a thing shameful, as they do now. Others disagree. Several women spoke of farming alongside their fathers and mothers when they were younger. It seems that at Wad al Abbas, as elsewhere in Sudan, the irrigated scheme reduced the role of women in agriculture (Bernal 1988a; Brausch 1964; Gaitskell 1959; Snyder et al. 1977). Nonetheless, older women and young girls, not subject to strict seclusion, do most of the cotton picking today, and a few women manage their own land and take part in other agricultural operations. While villagers disapprove of women working in agriculture or any public setting, in practice, women's activities are responsive to family circumstances. Moreover, evidence suggests that the ideal of female seclusion was less pronounced and less attainable in the past.

The importance of slave labor at Wad al Abbas is hard to assess. In 1905 the village was reported to have a population of 1,200 free villagers and 300 slaves (Gleichen 1905), which does not suggest an abundance of slave labor at the turn of the century. Villagers say that slave ownership was not universal, and that slave-owning households generally had three or four slaves. Some villagers were active in the slave trade. One of them, a wealthy man by village standards, owned 13 slaves. Family members and slaves worked side by side in farming and domestic chores. Family labor was the essential basis of agricultural production, but slave labor probably was a significant factor in differentiating large and small landowners. Land was not a constraint on agricultural expansion, but labor and the low level of technology were. Farm sizes in other villages apparently decreased with the end of slavery (McLoughlin 1962; Randall 1958).

The use of slave labor enabled some farmers to cultivate larger holdings, and perhaps to further diversify their economic activities by freeing some of their household labor from farmwork.

The slave trade began to decline under the Mahdi (Warburg 1981), and slavery was finally outlawed by the British in 1924. Upon gaining their freedom, the majority of slaves left Wad al Abbas since they had no land or means of support.

Cooperative labor parties *(nafiir)* offered another means of supplementing household labor. But Wad al Abbas farmers did not routinely rely on *nafiir* for agricultural labor. While in the Gezira poor farmers, before the irrigated scheme, apparently cultivated by *nafiir* (Barnett 1977), at Wad al Abbas *nafiir* were primarily called to farm the land of someone unable to cultivate his own fields (because of illness or some other circumstance), and to farm the land of the *faki* and the *khalwa* teachers. *Nafiir* also were used to build houses. Participants in a *nafiir* were usually given breakfast *(fatur)* by the landowner.

Nafiir may have been more important in the early decades of village history, about which knowledge is scant, but they played little role in farming in the period immediately preceding the establishment of the scheme at Wad al Abbas. Perhaps *nafiir* were eclipsed by hired labor once seasonal labor migrants were drawn to the central riverain region to work on the Gezira Scheme.

Family labor was the basis of household farming, however, and hired labor, though advantageous, was not an essential component of agricultural production. In addition to migrant laborers, local nomads were a limited source of seasonal wage-labor. The few freed slaves and their descendants who remained in Wad al Abbas also worked as paid laborers, and some villagers hired themselves out in addition to farming their own land. Hired labor was mainly a supplement to household labor at peak production periods, and was used by a few wealthier villagers rather than by the average farming household. The wealthier villagers were the more successful traders, religious leaders, and political figures such as the *nazir*, the *omda*, and the *shaykh*, whom villagers sometimes refer to collectively as *nass al kubaar* (literally, "big people" or "elders").

With the low level of technology and the household as the basic unit of production, class differences at Wad al Abbas did not arise primarily out of the agricultural base. Although land holdings were unequal, there was no appropriation of land from small farmers by large farmers. The abundance of land allowed some to expand without disenfranchising others.

Animal Husbandry

In addition to farming, villagers kept livestock. While villagers' accounts focus on cattle, smaller numbers of goats were probably kept, as they are today, as a source of milk and, to a lesser degree, meat. Villagers used donkeys for transportation and as beasts of burden. (Donkeys remain important means of transportation for shorter distances—within the village, to fields, and to neighboring villages.) Some villagers may have kept a camel or two for the same purposes, and for meat, as a few villagers still do.

Three main economic values are associated with livestock ownership in Wad al Abbas today, which were equally or more important in the past. First, livestock represent a source of food. Slaughtering for meat is limited, so as not to deplete herds. Dairy products, however, are a prized and important part of the diet. Milk, curdled milk *(rowb)*, and clarified butter *(semin)* are used in many foods, and were relied upon even more in the past when fewer items in the diet were purchased. Villagers say a family could live off these dairy products if need be. Second, livestock represent a store of wealth to be bartered or sold in emergencies. Third, livestock are productive assets in that they yield offspring.

Villagers were unsure of the size of herds in the past. Some reported that their grandfather's and great grandfather's herds were much larger than their own. Others said that in the past, people owned fewer cattle than villagers do now. The current distribution of cattle is very uneven; a few men own up to several hundred head, while many own no cattle at all.[3] About half the households in the village own cattle, usually between two and ten head. The contradictory statements by villagers could both be correct since it seems that, as a result of the scheme, herd sizes decreased even as livestock ownership became more widespread. While cash incomes from cotton production enabled poor farmers to buy livestock, the scheme eliminated pasture, making large herds expensive to maintain. Herd size declined for this reason and because there now are more attractive avenues of investment for wealthy merchants.

In sum, agriculture was the primary subsistence strategy at Wad al Abbas and trade was the main channel of accumulation. Animal husbandry played an ancillary role in terms of its contribution to subsistence and as an avenue of accumulation.

Crafts

Crafts have long been a subsidiary economic strategy at Wad al Abbas, but there do not appear to have been any full-time artisans in the village until recent decades. None of the villagers identified their early ancestors as having been craftsmen. Crafts and trade were always carried out in conjunction with farming. Hand-woven cotton cloth was produced in the village until the 1930s. Weaving may have been a carryover from the north, where it was widespread. Some women remember how to spin cotton, but there are no looms left in the village and weaving has disappeared from the region. The organization of production was based on a sexual division of labor—women spinning, men weaving—and was probably carried out at the household level. Cotton cloth *(dammuriya)* was produced for exchange as well as for use, as was the case in El Gedid (Randell 1958).

Traditional shoes *(merkoub)* have been made in the village since at least the late 1800s, and were sold in the village and in nearby markets like Wad Medani. As it is carried out today, *merkoub* production is an exclusively male activity, usually practiced on an individual basis. The same is true of the other crafts that have survived such as *angaraib* (traditional bed) making, and saddle *(sarj)* making. The level of skills in the crafts at Wad al Abbas today is rudimentary and goods are made to a standard pattern by all craftsmen. Although craft skills are often learned in the family, crafts are not inherited occupations.

Women continue to weave fiber food covers *(tabaqs)* for home use and for sale. They must have made other things for household use in the past. *Tabaq* production for sale, however, is part of the general commercialization of items of household consumption that has taken place over the last thirty or forty years.

Today few household utensils are produced by the household or even in the village; they are purchased from outside. As items of household consumption have become commodities, petty commodity production has gained in importance, and skills that were once widespread are now marketed by specialists. For example, houses are no longer constructed by *nafiir* (cooperative labor parties) but by paid builders. Crafts died out and skills were lost as imported manufactures replaced home-made goods.

Trade

Some of the earliest settlers at Wad al Abbas were traders as well as farmers. A small number of men, probably no more than ten, were slave traders. Villagers say slaves were brought from the *jibaal* (the mountains), and from Ethiopia across the border from Kurmuk. Villagers were not sure of the location of the mountains, but it is likely that some slaves were taken from the Ingessena hills further south in the Blue Nile Province. East of Wad al Abbas, along the Dinder and Rahad rivers to the Ethiopian border, was one area of Ja'ali slave raiding (McLoughlin 1962). Traders from Wad al Abbas may even have gone as far as the Nuba hills in Kordofan. Villagers say the slaves were sold to Egypt, but it is not clear whether traders from Wad al Abbas actually traveled there. Slaves were sold in the Wad al Abbas market and in other local markets.

Slave trading and other forms of long-distance trade were important sources of wealth. Long-distance trade was undertaken by villagers during the *saif* when there was no agricultural work to be done. There were several big traders in Wad al Abbas who advanced money and goods to other villagers to trade in Ethiopia, in return for 50% of the profits. One big trader was Hamad Said al Awad who grew sorghum, traded, and gave *shayl*. Under the Turks (1821–1885), when heavy taxes drove some northern Sudanese peasants out of agriculture into trade, they surmounted their lack of capital by pooling resources and trading collectively. Wad al Abbas villagers probably traded collectively as well.

Although trade flourished, the Sudan lacked a common medium of exchange until the British introduced currency (Sellin 1980:624). There were no coins under the Funj until the eighteenth century when the Spanish dollar was adopted in Sennar. Prior to that, gold and other items were used as currency (O'Fahey and Spaulding 1974:51, 52). Nor was there a standard medium of exchange under the *Turkiya*, though in some areas hand-woven cotton cloth was used and in others, such as eastern Sudan, salt (Ahmed 1977:33, 350), while *dura* (sorghum) was used as a medium of exchange along the Nile corridor (Issawi 1966:467). In the middle of the nineteenth century most of the exchanges at the trading center of Messelemiya in the Gezira were by barter (Hill 1970:29).

Villagers report that slaves were purchased in their region of origin in exchange for salt at one time, which is confirmed by McLoughlin (1962). Gold was used as a medium of exchange at Sennar under the Funj and under the *Turkiya*. Turkish and Ethiopian currency were also used in the area. Villagers referred to the Turkish money as *mediidi*, perhaps a corruption of *majidieh*, a silver coin of Turkish origin (Issawi 1966), though there was also a Mahdist coin known as *majiidi*. Villagers called the Ethiopian coins *rushali* and *gushiri*.

One trade route used by villagers led southeast from Wad al Abbas, across the Ethiopian border near Kurmuk, to Gondar and Asosa. Traders traveled by donkey and on foot, sometimes contracting with camel nomads to transport their goods. They took with them locally woven cotton cloth *(dammuriya)*, cotton cloth from Shendi, and salt, and brought back coffee, gold, and honey. A return trip took two to three months. This must have been an established trade route as Randell (1958:34) reports that merchants from El Gedid in the northern Gezira, also crossed the border at Kurmuk to trade for Ethiopian coffee and gold at Asosa.

Villagers traveled another trade route northeast to the Red Sea for salt, which was traded in the southwest, perhaps in southern Kordofan, for gold and slaves. Randell (1958) mentions an eastern trade route to Kassala and Gedaref where cloth was sold for sesame oil, honey, coffee, and dates, probably the first section of this northeast route. During the *Turkiya*, traders from Wad al Abbas also bought cowry shells and ivory in Sennar and sold them in Omdurman. The old man, Haj Hussein, says his father traded in chick peas, beans, squash, and salt which he sold to the Turks in Sennar, buying gold and ivory to sell elsewhere.

At the beginning of the nineteenth century, the Nile was not much used for travel in the Sudan; long-distance trade was carried out by camel caravan (Issawi 1966:464). After their conquest in 1821, the Turkish government launched boats connecting many areas via the Nile. However, caravans remained important until railroads were built by the British in the early twentieth century (Issawi 1966:468). Older villagers remember traveling by boat from Wad al Abbas to Wad Medani and Omdurman to sell grain. Some villagers also cut timber and floated it downstream for sale in Omdurman. Traders came by boat from those larger markets to buy grain directly from villagers.

The kinds of profits to be made in the grain trade from early on are indicated by the fact that, in the 1830s, the price of a 200-liter *ardeb* of sorghum rose from 6 piastres to 200 piastres (pt) in a short time (Hill 1970:53). Already at that time, speculation and hoarding were attractive, if risky, avenues of accumulation: "By forestalling grain a man may triple his capital or on the other hand ruin himself, for it is a risky business all right" (Hill 1970:177). An 1898 report quotes sorghum prices of 160pt/*ardeb* in Omdurman, 22pt/*ardeb* in Gedaref, and 48pt/*ardeb* in Kassala (in Issawi 1966:493). Such regional variation in prices allowed middlemen to realize great profits simply by purchasing grain in one town and selling it in another.

According to villagers, Wad al Abbas, which the Turks made a *markaz* (administrative outpost), was as big as Sennar during the *Turkiya*. From at least the early 1900s, a weekly market was held at Wad al Abbas (Gleichen 1905), which was one of the largest villages in the region, and served as a small commercial center for its area of the east bank *(Al Shariq)*. It was under British rule that present-day Sennar developed into a principal administrative and commercial center. (The Funj capital of Sennar, referred to by villagers as *Sennar al gadiima* (Old Sennar), began to decline in the eighteenth century (O'Fahey and Spaulding 1974) and must have been destroyed by the Turkish conquest or soon after.) Older villagers still call present-day Sennar *"Mukwaar,"* its original name meaning "the mixing together of different things," a reference to the ethnic diversity of its inhabitants.

Sennar *(Mukwaar)* began to grow with the influx of workers for the building of Sennar dam in 1914. It was then that a few traders from the north, such as Humeida from Berber, settled there and the market expanded to serve the laborers and administrators. Even before the development of Sennar *(Mukwaar)*, the nearby town of Sennar *Taqatu'* (Sennar Junction) grew up around the railroad that connected the area to Wad Medani and Khartoum in 1910 and to El Obeid in 1911. The Sennar dam, completed in 1924, bridged the Blue Nile at Sennar and made it much more accessible to Wad al Abbas and the rest of the east bank.

Until the late 1940s, however, villagers from Wad al Abbas continued to travel downstream by boat to Wad Medani and Omdurman markets to trade rather than to Sennar. It wasn't until the cotton boom of the 1950s and the proliferation of irrigated schemes throughout the area that Sennar's market really grew. Before that time, Sennar was smaller and transport difficult and time-consuming. Thus, for most of its history, Wad al Abbas was further removed from a major trade center than it is today; it was not so much Sennar's hinterland as a small regional center in its own right. But, once Sennar emerged as the primary trading place of the larger region, the commercial development of Wad al Abbas became linked with and subordinated to that of Sennar. Merchants from Wad al Abbas began to sell local produce to bigger merchants in Sennar. One of the wealthiest men in the village under the British, Hassan Beshir, acquired *dura* locally through *shayl* (crop mortgage) and purchase, and sold it to Humeida in Sennar. Hassan was the first villager to accumulate sufficient capital to buy a truck, probably around 1950.

MERCANTILE ACCUMULATION AND TRADITIONAL AGRICULTURE

Trade and agriculture were linked to each other at Wad al Abbas in several ways. A good crop could provide capital to enter trade. As one trader explained, "People here use up their grain so fast, giving it to their relatives. A trader keeps it and sells it." Another successful trader, explaining how some people accumulated wealth, said: "People start with farming, then they buy sugar, tea, soap. If a man can sell 150 or 200 sacks of grain then he has £S500 with which he gets goods on credit worth £S2,000 from a merchant. Then in the *khariif* [rainy season before the harvest] the things he bought for £S1, he sells for £S4." Wealth from trade, in turn, could be invested in agriculture. Even today if someone has surplus grain or cash crops to sell, farmers explain, "with this money he will buy charcoal and other things to trade and this will increase the money, then next agricultural season, he'll use the money to farm again."

The more successful traders were able to extend their landholdings using slave and, later, hired labor. In addition, these traders gained control over the crops of other peasants through *shayl* (crop mortgage). *Shayl* at Sennar dates from the eighteenth century when the Funj Sultan's power was declining and private merchants emerged out of the peasantry (O'Fahey and Spaulding 1974:81–82). By the early nineteenth century when Wad al Abbas was settled, *shayl* was thus an established practice in the region and probably operated in the village from its founding or soon thereafter. Traders acquired rights to farmers' crops cheaply through *shayl* and also purchased crops outright.

At El Gedid, downstream on the Blue Nile in the Gezira:

> It was the riverain crops from which a regular surplus could be accumulated that led to the use of commodities such as dried bamia [okra] and khadra [sic] [jew's mallow] as articles of trade. With this domestic source of supply as a basis for their enterprise merchants from the village expanded the scope of their operations and bought various goods from the markets of Omdurman. (Randell 1958:32)

Villagers at Wad al Abbas, however, emphasized the crucial role of sorghum in providing the basis of accumulation, as well as subsistence. Traders sold local grain in Medani and elsewhere downstream.

While Wad al Abbas had its own market from at least the turn of the century and was linked by long-distance trade networks to the market system, the average farmer remained geared toward subsistence production. Cash needs and dependence on the market appear to have remained limited into the 1940s. In a 1931 report by the English governor of the Blue Nile province, "the condition of the country outside the Gezira [Scheme] was described as a plethora of grain, meat, and milk but a complete lack of cash" (Gaitskell 1959:161). Cash was scarce at Wad al Abbas as evidenced by the reliance of villagers on barter arrangements for many things. There was a customary standard rate of exchange between grain and milk that operated in trading with nomadic herders who passed through the area, and *shayl* was largely a barter arrangement. Although as far back as men alive today remember, bridewealth was paid in cash, older men (including those who married as recently as the late 1940s) always speak of it in terms of the number of *ardeb*s of sorghum they had to sell to raise it. This suggests that local farmers did not commonly use substantial amounts of cash and that the sale of several *ardeb*'s worth of grain was restricted to exceptional expenses rather than a routine practice for the average farmer at Wad al Abbas. An old woman explained to me:

> In the past you dug holes, *matmura*s, and buried your grain and stored it til it went bad. People back then had no understanding of selling and it [grain] was so cheap anyway, they didn't sell it, just kept it in the *matmura* until it spoiled. But, in the past, spoiled *kisra* [flatbread], we'd eat it and spoiled grain, we'd put it in the fire and eat it.

While the wealthier families often had larger land holdings, trade, not agriculture, was the basis of economic differentiation. Tradewealth allowed some men to buy slaves or to hire labor and bring more land under cultivation. Wealth was accumulated primarily through trading and *shayl* rather than by producing crops. Eventually a small group of wealthy merchants began to emerge from the peasantry and form links with their like in Sennar and throughout the region. Part of the peasants' product was siphoned off through these trade links while local merchants accumulated wealth. The impact of this on agriculture, however, was restricted by the abundance of land and the slow development of the cash economy. The organization of agricultural production underwent little change and commodity production remained limited. Land did not become a commodity, and even while a market in agricultural labor emerged, most farmers relied on household rather than hired labor in cultivation.

The vast majority of villagers were farmers cultivating their own land to provision their households. Though land holdings were unequal, there were no landless and no landlords. As one man explained, "Everyone was a farmer. There were traders but they went on donkeys and sold little by little." Because of the important role played by long-distance and regional trade in accumulation at Wad al Abbas, local merchants extracted surplus from a large region. Their wealth was not simply based on exploiting fellow villagers, and their accumulation therefore was not associated with impoverishment of a corresponding magnitude among village farmers.

As trade goods became more available under the British, and household consumption of purchased goods rose, the dependence of villagers on *shayl* increased. Cash became more important, and subsistence production and petty commodity production overlapped as food crops were marketed when a surplus was produced or to meet pressing cash needs. Some farmers may have expanded sorghum production or added cash crops such as cotton and sesame. As the regional agricultural labor market grew, local farmers could hire workers in order to expand production, but few had the means to pursue this strategy. Until the mid-1950s when the scheme altered agricultural practices, farmers say they farmed with family labor, and only occasionally hired labor to supplement household labor at peak periods.

The relatively unchanged form of agricultural production at Wad al Abbas masked fundamental changes in peasant household economy, however. As the cycle of household reproduction came to include the merchant as a necessary link, providing goods and credit, farmers became dependent on capital. Thus, while they retained control over their farms, market conditions beyond their control played an increasingly determinant role in their destinies. These changes appear to have been gradual, however, and the subsistence base of the village economy remained viable into the 1950s.

Prior to the scheme, there was almost no labor migration from Wad al Abbas. A few men went to work on the railroad at Sennar, others worked on the Sennar dam, and at least one villager served in the army under the British. But, as one man put it: "In the past, people didn't travel. You weed your fields and you come home." There was little occupational specialization in Wad al Abbas; traders and craftsmen remained farmers as well. A boy farmed with his father, grew up, married, and eventually farmed his own land. Traders traveled and returned. Farming and, to a lesser degree, trade were the basis of the village economy into which new generations were absorbed.

In 1954 the establishment of an irrigated cotton scheme by merchants from outside the village spurred rapid changes in village economy as compared to the evidence of continuity throughout its earlier history.

IRRIGATED AGRICULTURE

Private Management

Wad al Abbas is located in the main cotton-growing region of Sudan where the waters of the Blue Nile are used to irrigate the clay soil of the surrounding plains. Under license from the British colonial government, a pump scheme was established there in 1954 by two merchants from outside the village, Abdel Rahim Ahmed al Terzi, originally from Berber but residing in Wad Medani, and Abdel Hamid al Mahdi of Omdurman.

Al Tayeb Mohammed Abdel Rahman Ab Ziig, the *Omda* of the region from 1942 until the abolition of Native Administration in 1972, was born in Wad al Abbas where he resided until his recent death. He said that he entered into a partnership with Abdel Rahim under which 4,500 *feddans* were registered in his name. Abdel Rahim then allotted 300 tenancies of 15 *feddans* to villagers and managed the scheme. The cotton crop was split between the two men and farmers received a share of the proceeds once it was sold. At that time the *Omda* was an influential figure and wealthy by village standards, with large rainland holdings and livestock. Osman (1958) sheds light on why the merchant would have agreed to such an arrangement. According to him, capital was available in the 1950s but obtaining a license for a cotton scheme was difficult. "To be eligible for a license, that is to be a man of some influence in one of the areas of the Province, was sufficient to guarantee a 50% share in the net profits of a scheme" (Osman 1958:45).

The scheme at Wad al Abbas incorporated nearly all the land villagers had traditionally cultivated by rainfed agriculture. Villagers received tenancies as their compensation for this expropriation of their land. One villager remembers the establishment of the scheme with some bitterness:

> They didn't buy the land. They said it belonged to the government. No one talked to them, they just came and took the land—without any compensation. They said, "we will give you tenancies." People said, "give us money for the land." But they said, "No, we will give you tenancies."

There were enough tenancies for every man who wanted one to have it, so no group of landless was created by the scheme. Individual holdings were limited to no more than two 15 *feddan* (6.3 ha) tenancies, but those who had had larger holdings of rainland were compensated by being allowed to nominate other tenants. These men, through nominating infant sons or other such means, effectively gained for themselves ownership of several tenancies. Thus, from the start, farmers' holdings on the scheme were not equal, but reflected pre-scheme inequalities. According to a villager's astute analysis, this was a way of silencing any potential opposition from wealthier villagers. Women were not allocated tenancies although they had owned land prior to the scheme (Bernal 1988a). It is likely that male relatives included women's holdings in their own land claims. Female landowners were thus completely expropriated. The few women who own tenancies today gained their land through inheritance.

Scheme regulations require farmers to have one third of their land under cotton and another third fallow each year in yearly rotation. Most farmers choose to grow sorghum on the remaining third of their land though some occasionally plant peanuts or coriander. Until 1981, the scheme rented the land from the tenant (for a nominal sum), not the other way around. In any case, scheme management controls the irrigation system, regulates land use, and ultimately has the power to reallocate land from a farmer deemed unproductive to another. Farmers' land rights, thus, are limited in practice. It is illegal for farmers to sell tenancies although they may transfer ownership to someone of their choice. Tenancies are inherited by the owner's heirs and farmers feel the land to be theirs. A woman tenant reported the following conversation she had with a scheme employee who wanted to revoke her tenancy because no cotton had been planted on it: "I met one [of the inspectors] and I said to him, 'you come here now and you will take from me what I have had from long ago, what I was born on?' 'Who is your *samad* [village inspector]?' he said to me. 'Allah,' I said."

Cotton production on the irrigated schemes was conceived by the colonial government in terms of a profit-sharing partnership between tenants and management. Management would supply irrigation and other inputs, as well as gin and market the crop. Tenants would carry out the cultivation and harvesting of cotton, and receive a share of the proceeds when the crop was sold. A Wad al Abbas tenant who remembers the establishment of the scheme says:

In those days the traders [who owned the scheme] made loans to you. They give you money to weed and pick and then take it from the cotton. Then, too, the sorghum was your own like now; they only own the cotton. But, he'd give you a little [money] to eat and say, "that's all," but that wasn't all. He'd eaten it himself.

The notion of partnership misrepresents the degree to which the organization of the scheme placed cultivators in a position subordinate to the elites that controlled it. The scheme did not convert farmers into landless workers. But it weakened farmers' hold on their means of production, reduced their control over the production process and over their products, and brought local resources under the control of the state and a mercantile capitalist class to serve "national interests" rather than to meet local needs (Bernal 1988b).

The system of irrigated agriculture requires much more labor than the traditional rainfed system of cultivation and the agricultural cycle is longer. An older farmer describing the agricultural cycle before the scheme, lamented, "But now the whole year, there's not one day free." Men and boys do most of the farming whether as unpaid labor or hired laborers. Few women own land or farm as part of the family labor force. Women's role in scheme agriculture is limited to harvesting cotton which, unlike other operations, is almost exclusively carried out by paid workers. Even a daughter picking her father's cotton is paid. Old women and young girls do much of this arduous, low-paying work. The scheme thus reduced women's control over land and their participation in family farming, while drawing them into seasonal agricultural wage-labor. The official production schedule on the scheme was (and remains) as follows:

April–May	Fields which had been under cotton the preceding year are cleared in preparation for planting sorghum.
June–July	Fields which had been fallow the preceding year are cleared in preparation for planting cotton.
June 15th	Water is opened on the sorghum land.
July	Sorghum is planted, cotton land is split-ridged by tractor, and water is opened on the cotton land.
July 15–August	Cotton is planted, and after 21 days it is thinned; both cotton and sorghum are weeded.
September–October	Cotton and sorghum are weeded.
November–December	Sorghum is harvested and threshed in the fields.

January–March	Cotton is picked several times. After the final picking, cotton plants are pulled out of the ground to reduce the spread of disease. The clearing of fields begins for the next agricultural year.

Despite the hardships of the new system, farmers remember the early years of the scheme as profitable. As one stated, "In the beginning the scheme gave a big yield, each tenancy produced forty or fifty *kantar* or more. Now there's no production and no profit." Another said, "Before, we used to get six *kantar* of cotton from one *feddan*, now if we get three or four, it is considered good." A third reminisces, "In the past, even if you produced a little you got paid." In the 1950s, cotton profits brought new prosperity to the village, credit offered by the scheme resulted in the decline of *shayl*, and the standard of living rose. Before the scheme, all but a few wealthy men lived in huts (*gotiyas*); cotton money allowed villagers to construct more substantial square dwellings, of earth or even brick. Cotton profits also stimulated village self-help projects which constructed schools and other public buildings, such as the health and veterinary stations. Most of these buildings still serve today.

The scheme brought unprecedented amounts of money into villagers' hands, drawing them into the cash economy. While prior to the scheme farmers had sold some of their rainfed crops to raise cash for specific purposes like bridewealth and for the purchase of necessities such as tea and clothing, irrigated cotton production made cash more available, spawning new consumption patterns. Scheme land-use regulations also meant that farmers no longer grew a wide variety of subsistence crops— they grew only their staple grain, sorghum—and therefore had to buy many items of their diet. Purchased goods began to assume an essential role in household economy, and villagers grew dependent on the flow of cash to survive, and to maintain their new standard of living. Cash also became a vital input to farming as hired labor was necessary to meet the demands of the new production system, particularly for harvesting cotton. Gaitskell (1959) has described how this situation led to increased consumption, inflation, and ultimately to debt for Gezira tenants.

The new prosperity at Wad al Abbas was short-lived. In the 1960s, world cotton prices plummetted. Local cotton profits and productivity fell, and inflation decreased the real value of monetary incomes. In 1966, angry Wad al Abbas farmers stormed the scheme office and attacked the scheme owner and some employees, demanding better payment for their cotton. The farmers' revolt was put down by the police and the army. A number of Wad al Abbas men were tried for their actions and sentenced to years in prison, but the sentences ultimately were reduced to several months. A farmer who participated in the uprising remembers it this way:

> All the farmers got together here because the pay [for cotton] wasn't good. We went down and took all the people in the office [of the scheme] and the owner, too, and beat them. The police and the army were brought in from Medani and Sennar. They arrested twenty-one men and put them in jail. They fired tear gas to disperse the people.

"Since then, there has been no strike here." he adds. The scheme was nationalized two years later and conditions continued to worsen for Wad al Abbas farmers.

State Management

Along with most of the private cotton schemes, the Wad al Abbas scheme was nationalized in 1968 and is now administered, as part of the Blue Nile Schemes, by a government corporation, the Blue Nile Agricultural Corporation *(al Mu'asessa al Ziraiyya lil Nil al Azraq)*. Nationalization did not alter the organization of production or the subordinate position of farmers, and the physical infrastructure of the scheme deteriorated. Moreover, costs of production on the pump schemes rose rapidly after nationalization, mainly due to overstaffing and increasing fertilizer costs (Oesterdiekhof 1980a:307).

Nationalization brought Wad al Abbas farmers under a large national bureaucracy. Scheme administrative units do not correspond to villages. There are three administrative divisions in the scheme at Wad al Abbas, each of which includes farmers from other villages as well as from Wad al Abbas. The two scheme offices near the village, Wad al Abbas and Al Saar, also administer several other Blue Nile Schemes in the area, including those at Shambada, Gheresli, and Wad Tawil. These were established as separate private schemes and later brought under central management with the scheme at Wad al Abbas. The next higher regional administrative office for farmers at Wad al Abbas is at Kassab, closer to Sennar but still on the east bank of the Blue Nile. The main offices are across the river in Sennar. Ultimately policy is set by the Ministry of Agriculture in Khartoum. One farmer had this to say about nationalization:

[The scheme] became government and the government says farmers can't strike [refuse to grow cotton]. If there is a problem they should complain to the head office in Sennar. If that doesn't work, they should go through the [tenants'] Union[4] and people from the Union will see the matter in Khartoum, see whether the complaint is justified or not. . . . The people in the office now are not responsible. They are just employees, all the responsibility is in Khartoum. Even in Sennar if you tell him "I won't farm" he says, "As you like, it's not my affair." They are secure, it doesn't matter whether the farming goes well or not, they get paid.

Until 1981, farmers continued to produce cotton under a profit-sharing arrangement with management. Any other crops were the farmer's alone and management provided no inputs, credit, or extension work to support their production. Initially, cotton profits were divided 40/60 between tenants and management; by 1980 farmers' share was 50 percent. Under this system any costs associated with inputs aimed at improved yields or marketing were deducted from a "joint account" *(hisaab mushtarak* or *hisaab al ishtiraqi)* and shared by tenants and management. As a result of the agitation of farmers throughout the schemes, particularly on the Gezira, items were added to the joint account over time so that by the 1980s it included costs of seeds, fertilizers, aerial pesticide spraying, plowing, transportation, advances to tenants, ginning, storage, and insurance. Whatever remained of cotton proceeds after these deductions was then subject to the profit-sharing division. Some farmers felt this penalized the more productive since the proceeds from their yields were used to cover the debts of others in the joint account before profits were calculated.

Profits and productivity stagnated through the 1970s while costs of production climbed. Between 1976 and 1980 the price of fertilizer doubled and fuel costs for the pumps increased (Oesterdiekhof 1980a). The head administrator of the Wad al Abbas scheme office reported the average yield for 1979–80 as only 1.5 *kantar/feddan*, an extremely low yield even by Sudanese standards. Scheme personnel and tenants alike reported that 1979–80 was no aberration, but typical of conditions there.

By 1980 most Wad al Abbas farmers hadn't received any cotton profits in years. Of 1,597 tenants under the administration of the Wad al Abbas office 1,591 were in debt to the scheme for production costs and credit. Debts accumulate on a tenant's account from one year to the next, to be deducted from his future cotton profits. Farmers are never required to pay these debts from other sources but they risk losing their land if they are consistently poor cotton producers. Cumulative debts also make it harder for a farmer to make a profit even in a good year. Many farmers actually look to credit advances as their only income from farming: "A man gets only the credit for the work," says a farmer's wife. Farmers received about £S136 (US$170) credit per year for a tenancy in the early eighties.

In 1981 the joint account system was replaced on most of Sudan's schemes, including the one at Wad al Abbas, by an individual accounting system whereby farmers are charged for land, water, and services provided by management and are paid a fixed price for their cotton. Charges for land and water were extended to all other crops as well as to cotton. One farmer's reaction to this was typical, "Now they want to charge for the water to sorghum and the water to coriander. Where will a farmer get such money?" Actually charges are calculated in terms of a minimum cotton yield (about 3 *kantar/feddan*; 321.3 kg/ha) for which farmers receive no payment. Farmers are held in debt to the scheme if they fail to produce enough cotton to cover this. In practice, individual accounts made little difference for most Wad al Abbas farmers, whose yields fell below the minimum. Other features of the scheme remain unchanged.

The farmer's role in cotton production is to supply labor. The farmer is responsible for the sowing, weeding, picking, and pulling out of cotton plants. This can be done by family members, sharecroppers, or hired laborers. Tenants generally employ hired labor for some of the cotton work and the scheme provides credit to help them meet their labor bills. These advances are added to the production costs farmers must cover before receiving any payment for their cotton.

Relations between scheme employees and farmers are not cordial. With the exception of a few wealthy and powerful villagers, scheme employees do not visit villagers' homes. When necessary they find farmers in the village market. One employee explained, "We're from Khartoum. We consider Wad al Abbas bad country." While scheme management tends to view farmers as lazy and recalcitrant and holds them responsible for poor yields, farmers have similar complaints about management. As one farmer expressed common sentiments: "There is no work in that office, it is empty administration, that's all."

The basic characteristics of irrigated agriculture at Wad al Abbas are: compulsory cotton cultivation under conditions largely determined by management, the absence of a free market in land, monocrop subsistence production alongside monocrop export production, and low yields in both major crops. A closer examination of cotton and sorghum production in the 1980s reveals the dire circumstances of Wad al Abbas farmers. Agriculture is no longer a reliable source of subsistence or income.

Cotton—Where's the Cash in Cash Crop Production? Perhaps a farmer put it best when he declared: "I will tell you so you understand about this cotton. 'Farmer' means death!" He went on to explain, "I go to the island in the river and cut thornbrush, and sell it for 10pt and 20pt to feed my children. My name is 'farmer.' This cotton, you go and work on it all year and finally they tell you you're in debt."

The major reasons for the lack of profits to farmers are: low cotton yields, high production costs, and the method of determining and allocating costs and profits between tenants and management. World cotton prices are also closely related to the profitability of cotton cultivation on the scheme. Low yields and high production costs are relative to the price at which cotton can be sold on the world market, over which Sudan has little control. While many cotton schemes were established in the 1950s when world cotton prices hit record highs, rarely have prices reached those levels since.

The reasons for low yields at Wad al Abbas include crop diseases such as white fly (locally called *asel* [honey] because it leaves a sticky substance on plants), technical breakdowns, livestock trespass, water shortages, and probably declining soil fertility. Irrigation equipment and canals were not properly maintained after the scheme was nationalized. In the early 1980s, the foremost concern of farmers was the paucity of irrigation and the unreliability of its delivery. One farmer says:

> When the fuel for the pumps comes, [the scheme employees] don't put it in the pumps. They sell it on the black market. So the crops aren't good. There isn't enough fuel to run the irrigation pumps. The farmers complained to the ministers about this and the Minister of Agriculture has come to Wad al Abbas three times. He set up a twelve-member board to investigate. As a result, this year is better than the past.

These specific allegations are difficult to document, but they are consistent with widespread reports of corruption in many areas of public administration in Sudan.

Livestock damage to crops is a recurrent problem because the schemes in the area cover land that was used by nomads and local farmers to graze their herds. As one villager explains:

> There is no grazing land for cattle. Because of the lack of food for them, at night people let their animals loose to eat the things of the

farm. So you work all day in your field and at night, when you're asleep, cows come and eat everything. When you go in the morning you find nothing—no cows, no crops, nothing.

Low yields also result from the responses of farmers to the structure of the scheme and to the poor remuneration for cotton production. Farmers are excluded from many decisions regarding cotton production, and are required to keep another third of their land fallow. Farmers have no say over inputs to cotton production such as insecticide spraying or mechanization, nor do they have a voice in how or where their cotton is ginned, graded, and sold. Farmers thus have little control over production costs, although such costs directly affect their returns from cotton production. In view of these circumstances, it is not surprising that many Wad al Abbas farmers neglect cotton, which they generally see as a net loss. Farmers on Sudan's other schemes commonly do the same. Khalafalla (1981a:127–128) argues that when cotton profits declined in the 1960s, "tenants passively resisted exploitation by giving less attention to cotton production." Barnett (1981:323) reports that on the Gezira, "The tenants' consciousness of their situation effectively lowers the level of cotton production as they decide that labour expended on millet has a higher return than labour expended on cotton." By the early 1980s many Wad al Abbas farmers had lost hope of ever being paid for cotton and resented being forced to cultivate it. As one commented, "Even those who produce a lot get nothing for it. . . . [The scheme] doesn't give the farmers a thing. Not a penny."

On January 11, 1982, the school boys of Wad al Abbas rioted. Riots, strikes, and demonstrations had been going on in towns all over the country, intermittently, for a year to protest rising food and fuel prices and to express a lack of confidence in the Numeiri regime. Villagers were well aware of these events through word of mouth and through listening to "Voice of Libya" radio broadcasts addressed to the Sudanese people. Like townspeople, villagers had been suffering the effects of inflation, black market prices, and shortages of essential consumption items such as flour and sugar. A round of mass demonstrations and strikes erupted in Sudan that January because of IMF-imposed austerity measures lifting government subsidies of basic goods, particularly sugar. The village boys were thus participating in national politics, but their local targets are significant. They first ran to the village market. Shortages, inflation, hoarding, and black marketeering had put severe pressure on household consumption. The boys did some damage, throwing stones at shops and pulling a thatched roof down. But local traders know the boys and are their elders and successfully chased them off. The boys then assaulted the offices of the *Mejlis al Shaabi* (People's, [i.e. Government's] Council), inflicting minor damage to the building. Finally, they headed for the irrigation scheme office where they ransacked everything. (The staff withdrew to their government houses nearby.) The boys dumped out files and administrative records which blew in the dust and caught in thornbushes. They found paint and poured it on the buildings, furniture, and equipment, set fires in the offices, and burned a tractor.

This revolt, like the one in 1966 and others around the country, was met with repression. Local police and reinforcements from Sennar caught some of the boys and learned the identities of others. Officers went around at night to the children's homes and fearful parents felt compelled to hand them over. About thirty-seven boys were arrested and taken to Sennar jail where they spent several days awaiting trial. Due to their youth and the intervention of wealthy villagers with contacts in Sennar, the boys were sentenced to flogging and fines rather than jail. Such whippings are harsh punishment, however, administered severely with each stroke of the lash exacting great pain.[5] To my knowledge no investigation of any kind was ever conducted into the grievances that underlay this protest nor were any changes implemented as a result of it. The village returned to calm. As one villager put it, "We're all *ta'baan* [lit. tired but also means poor]. But we keep quiet." A returning labor migrant later reported that news of the boys' riot had reached Wad al Abbas men in Saudi Arabia.

In the mid 1980s, conditions on the scheme were ameliorated somewhat by the installation of a new pump which improved the water supply. Average cotton yields rose from around 1.5 *kantar/feddan* for the years 1980 to 1983 to around 4 *kantar/feddan* between 1984 and 1986 (BNAPC 1987). Even yields of 5 *kantar/feddan*, are low by world standards, however (ILO 1976:258). The percentage of tenants on the Wad al Abbas Scheme who received any payments for cotton production rose from 0.4 percent in 1979–80 to 38.3 percent in 1984–85 (BNAPC 1987). Even so, close to two thirds of the farmers continued to operate in the red. Moreover, farmers sometimes spend their own money on hired labor and therefore actually run at a loss although the scheme pays them some cotton profits. More light is shed on these officially recorded gains by a villager who explained to me that some tenants had secretly purchased cotton from others so as to inflate their own yields and receive cotton profits. (Some farmers are willing to sell cotton since, if a farmer's yield is below the official break-even point, his cotton is essentially worthless to him.)

In 1988 despite the improvements, Wad al Abbas farmers continued to express frustration and hopelessness about cotton production. Echoing the complaints of farmers familiar to me from the early 1980s, one said: "There's nothing in farming. You plant and you weed and then they make the accounts and they tell you good-bye." Moreover, farmers consistently reported that water shortage remained a major constraint on the production of all crops, despite the new pump, because the fuel supply was unreliable.

A rehabilitation program is now underway throughout Sudan's irrigated schemes and is expected to reach Wad al Abbas in the 1990s. The plan deals with some of the scheme's technical problems but does not alter the structure of the scheme or the relationship between farmers and management. It is unlikely therefore to significantly improve conditions for farmers. Irrigated land for their food crop, sorghum, is the sole benefit most farmers receive for participating in the scheme. As the daughter of a woman tenant explains, "My mother isn't interested in the cotton, but if she doesn't grow it, she won't have the land to grow sorghum."

Sorghum—Grain Is Worth More Than Money. Sorghum production has been institutionalized on the Wad al Abbas scheme from the beginning. It is a key to understanding the operation of the scheme in spite of low cotton payments. Farmers speak of cotton as *"haq al hukuma"* (belonging to the government); they speak of sorghum, on the other hand, as belonging to them. Farmers control all decisions regarding production on the third of their land they can devote to sorghum, and management has no share in the produce. Farmers can grow crops other than sorghum if they choose, but few do. According to villagers, when cotton profits were good, it was more common for them to grow peanuts as a cash crop, but the decline of income from cotton led to a return to subsistence farming.

The incorporation of subsistence production alongside export production on the scheme serves two major functions. First, it takes advantage of what has been called the "tributary function" of subsistence agriculture in peripheral capitalism (Oesterdiekhoff 1980b). That is that labor which produces all or part of its own subsistence can be paid less than its actual value (the cost of its reproduction). Subsistence production on the scheme contributes to the reproduction of the tenant population and therefore subsidizes cotton production, since farmers do not have to live on cotton profits alone. Second, by providing farmers with irrigated land for subsistence production, the scheme strengthens its control over them since farmers forfeit this land if they refuse to cultivate cotton. Not only must a farmer maintain a certain level of cotton cultivation to retain his tenancy, but cotton cultivation is monitored and a farmer is given the cash credit for each cotton operation only after an inspector certifies its completion. This limits the labor farmers can devote to sorghum or to off-farm work at the expense of cotton. Sometimes, for lack of money or labor, a farmer cultivates only cotton, simply to protect his land rights.

Despite the importance villagers accord sorghum, the yields of most farmers are below their consumption needs. Villagers depend heavily on the market even for their staple grain. In a sub-sample of 19 farmers, 53 percent produced less than half the sorghum their household consumed in 1980–81, while only 21 percent were self-sufficient in sorghum (table 3.1). In 1981–82, 42 percent produced less than half the grain they consumed, while 32 percent attained self-sufficiency in sorghum.

Farmers say a yield of six sacks of sorghum per *feddan* is possible in a good year, but only one farmer in the sample achieved this either year. Sorghum yields are affected by water shortage as well as the predations of livestock and birds. Moreover, because only one third of their holdings can be under sorghum each year, some households lack sufficient land to meet their sorghum needs. Elsewhere I have shown in more detail how the demands of cotton production on farmers' land, labor, and cash resources interfere with their efforts to produce food (Bernal 1990).

One indication of the poor returns from agriculture is the low value of land. It is not unusual for a wealthy man simply to give his tenancy

TABLE 3.1 Degree of Self-Sufficiency in Sorghum Production

Yield as a Percentage Of Consumption	1980–81			1981–82		
	#	Farmers	%	#	Farmers	%
100 or more	4		21	6		32
50–99	5		26	5		26
0–49	10		53	8		42
	N = 19		100%	N = 19		100%

NOTES

The 19 farmers represent 21 households both years due to the multiple households of polygynous men.

Calculations are based on each household's reported grain consumption and yields.

Two farmers in 1980–81 and one in 1981–82 employed sharecroppers who would normally receive half the crop; the entire yield is included here, however.

outright to a poor relation. A desirable tenancy near the village and with good irrigation sold (illegally)[6] for only £S100 or so in 1982, while one distant from the village and far down the irrigation line could go for as little as £S10. An average cow cost £S300 at that time.

The irrigation scheme effectively brought the peasants of Wad al Abbas under the control of the state and Sudan's fledgling capitalist class. While the peasants did not lose their land they lost much of their economic autonomy as they were forced into cash crop production. Moreover, the state, by setting the terms under which cotton is produced and delivered, is able to determine the level of returns to farmers. The unprofitability of cotton production under this system has limited agriculture as a means of accumulation. Subsistence production thus remains the primary goal of Wad al Abbas farmers. However, conditions on the scheme have undermined their ability to produce sufficient food for their families. Unable to secure household reproduction through cash cropping or through subsistence production, villagers turned to petty trade and wage-labor as sources of income.

The irrigation scheme thus contributed to the formation of the peasant-worker class at Wad al Abbas by eroding the peasants' capacity to sustain themselves solely through household agricultural production. In other areas this is brought about by declining farm sizes due to land concentration and/or fragmentation, ecological degradation, unfavorable terms of trade between agriculture and industry, taxation, the indebtedness of farmers to moneylenders or other creditors, and development projects which impose high production costs and/or risks without sufficient returns to farmers.

THE EXPANSION OF OFF-FARM WORK

Through the 1960s and '70s, as returns from agriculture fell, villagers were propelled into trade and wage-labor as survival strategies for the many and as avenues of accumulation for the few. Villagers themselves cite declining agricultural conditions as the main reason for the beginning of labor migration from Wad al Abbas. One says,

> I worked here fifteen to twenty years as a farmer, then Numeiri came and said the cotton wasn't good enough. Production wasn't good so there was no money at the end of the year. It's better for a person to

go out and work because the children need a lot. I work in Port Sudan in trade and send them money every so often.

Another man tells it this way:

In 1969 the government took over the scheme and production fell. Before, the traders [who owned the scheme] looked after everything themselves. Now, the government just sends civil servants. They are not responsible and the work became bad, not because of the farmers. The production fell. The farmers said, "You don't provide spare parts, fuel, water, etc., why should I grow cotton?" They would rather grow sorghum, peanuts, and coriander. [Farmers] get all the profits from these crops while the government takes the cotton. The youth left to work in trade in Sennar, Medani, Khartoum. They send the money back to their families to help them. Here, everything is expensive so all the youth go—to send money to their families to live.

Trade and transport are the most widespread activities among villagers because opportunities for wage employment are limited.

Trade

The inflow of cash from cotton schemes throughout the region during the boom years of the 1950s spurred the expansion of the village market at Wad al Abbas. The establishment of the scheme also determined that trade, not agriculture, would remain the primary avenue of accumulation for villagers. Though owners of larger rainfed holdings gained additional tenancies, scheme regulations set limits on the concentration of land and on the profitability of farming. In the initial years of the scheme, cotton profits provided villagers with capital to begin trade ventures, and simultaneously created a local demand for consumer goods.

In the early 1960s, business was good enough that some Wad al Abbas traders built permanent shops and storehouses in the marketplace, which until then had consisted only of makeshift stalls and mats on the ground. The variety of available goods increased. One villager bought a bus and ran it from the village to Sennar. A few local merchants were able to purchase tractors which they rented out to farmers to prepare land for sorghum cultivation and sometimes to the scheme for cotton operations.

It wasn't until the early '70s, however, that local mercantile accumulation really took off. Most of the wealthy merchants in the village today accumulated their capital in the 1970s. Villagers remember the early '70s as a period of great inflation, when prices of household consumption items and livestock began to rise rapidly and a thriving black market in government price-controlled commodities emerged. One villager relates this to national events as follows:

> All these shortages and the black market and inflation came from the beginning of Numeiri's regime when people were against him and he killed so many people. At that time, everything came from outside—sugar, oil, cloth, and other things. The opposition tried to create shortages to make him [Numeiri] fail and cause the people to strike. Then, when people finally found one of the things that were in shortage, they were willing to pay high prices. So, the merchants learned from this to get profits by creating shortages and selling goods on the black market.

In any case, certain men at Wad al Abbas were able to amass considerable trade capital during that period. Some of them accumulated capital by trading in distant regions of the country such as Kordofan, Dar Fur, and southern Sudan, joining the ranks of the *jellaba* (Ja'aliyiin and other northerners who monopolize trade links throughout the country). Their profits were primarily reinvested in trade to expand and diversify their commercial enterprises, rather than invested in their village farms. A few merchants purchased lorries to transport their goods and/or to establish freight businesses.

This growing prosperity was not evenly distributed. With the decline of profits from cotton cultivation, small food harvests, and the rising cost of living, many less fortunate villagers were pressured to enter petty trade to support their families. The majority of traders, big and small, deal in basic household supplies like tea, spices, oil, and soap, or in agricultural produce; what differentiates them is volume. Petty traders buy a sack of onions or some other staple and sell it in smaller units, eking out a living this way, but little more.

In the mid-1970s, the market at Wad al Abbas, which had been growing in the volume and diversity of its goods and the wealth of its merchants, began to decline and the center of local trade shifted to Sennar. It is not clear which came first—the shift of Wad al Abbas merchants away from their own village as the base of their enterprises, or the shift of locals throughout the east bank from Wad al Abbas to Sennar as the regional consumers market. Either movement could have stimulated the other.

Certainly transport had been improving. Whereas in the past, it would have been impractical for people to travel frequently to Sennar, there were now buses (converted trucks) running along the east bank and across to Sennar. About a dozen buses were owned by Wad al Abbas villagers and some other villages in the region had experienced a similar process of mercantile development and expansion into the transport business. It is also likely that, as Wad al Abbas traders accumulated capital and were able to deal in larger volume and in more expensive goods, local buying power simply could not support them and they went elsewhere to trade. Transporters similarly found they could make more money going from Sennar to Medani and other routes than from the village, and some of them also moved out.

By 1980 traders from Wad al Abbas had stores in Sennar, Khartoum, Omdurman, Kosti, al Demaziin, al Renk, and other towns. Some who made their fortunes trading out west in Kordofan returned with capital to establish stores closer to home. Some eventually relocated their families and are no longer part of the village, although they keep in touch with their relatives and friends there. A few men married second wives in the towns in which they trade. But most keep their households in the village where their wives and children are surrounded by close relatives, the cost of living is lower, and they can maintain farms and livestock.

A tenant who is now a full-time trader and commutes daily to his shop in Sennar says:

> In the past you wouldn't find a soul from Wad al Abbas outside. They were all here farming and the *suq* [marketplace] was good. Now they've all gone out. The reason for that is the agriculture here. The scheme doesn't benefit the farmer anything. The traders, too, have moved to Sennar and other places. Now there can't be fewer than 300 or 400 sons of Wad al Abbas outside.

The accumulation of trade-wealth at Wad al Abbas meant greater occupational specialization and social differentiation. Successful merchants explain their prosperity by contrasting themselves with poor people. As one puts it, "People who are traders don't just give people something for nothing [to help them]. They make them work for it, giving them money [on credit] to engage in trade." Such arrangements usually entail the repayment of principal plus a 50 percent share of profits. Like *shayl*, profit-sharing does not break the Islamic prohibition against interest. Obligations of mutual aid to community and kin thus were not so much abrogated by the wealthy as subtly converted into profit-oriented transactions. For these successful men the accumulation of wealth has taken on a moral quality so that those less fortunate are by implication foolish or profligate. Another explains,

> A poor person will sell and slaughter [cattle], but a rich person won't. . . . A cow gives milk which is expensive, and yogurt and butter. The cow is like his son, he loves it. He won't sell it unless he wants to buy a car or for a marriage, and the cows increase. He can sell the milk and yogurt and the bulls as they don't give milk or give birth. You sell the ones long in the tooth and buy young ones . . . But if he sells them all, his brothers will ask for money and there will be nothing left for trade.

In reality, pressed by immediate subsistence needs, most villagers are not able to accumulate wealth through their petty commercial activities. Like the proverbial poor man described above who must deplete his herds to live, the petty trader continually spends his meagre profits on basic household consumption and therefore is never able to amass seed money to improve his economic position.

Wage-Labor

There was never much of a market for villagers' labor at Wad al Abbas. While some villagers tried to earn a little cash as casual agricultural laborers for their fellows, such work was poorly paid and intermittent. Moreover, most villagers were short of labor on their own land during the peak periods of labor demand. Government institutions like the schools, and the scheme and council offices, brought a few jobs such as janitor or guard. There are also teaching positions in the local schools, but only recently have villagers been qualified for these. Generally, wage or salaried employment for villagers has meant migration. As one villager notes with regret, "There's no work here. Anyone who is good has left. Here there's only farming or buying and selling."

While petty trade can be carried out in the village or with occasional trips and can be tailored to the demands of farming, obtaining wage-work necessitates leaving home and farm. The falling cotton profits and inflation of the late 1960s made villagers ever more dependent on off-farm incomes, however, and by 1970 men were seeking wage-work in urban areas such as Wad Medani and Khartoum.

As one man laments,

> Before, unmarried boys worked with their father on his land and there was no employment then, everyone stayed. Now everyone has gone, leaving the house empty.

On second thought he adds,

> It is better now than before because before you may have had five or six family members to feed and you are responsible. They look to you. Now your sons, even before they marry, they work and take care of you.

Another villager says,

> You think this village is big, but there are so many people outside. You don't see them. They come for the *aeed* [Islamic holidays] and vacation and go back.

Not only did returns from agriculture decline, there was little new land available for younger men who had not received tenancies when the scheme was established. Few households are landless, but holdings have been fragmented through inheritance so that greater numbers of people depend on smaller pieces of land.[7]

While agricultural conditions forced villagers to seek other sources of income, the schools that had been established at Wad al Abbas in the 1950s began to produce some educated men. These men were qualified for clerical and semi-skilled jobs, such as driver, so wage-labor for villagers was no longer synonymous with low-paying, manual labor, but with more desirable jobs. The growing participation of villagers in wage-labor, since the 1970s, is thus partly a reflection of the increased educational opportunities in Wad al Abbas which qualify villagers for such work. But, villagers have been pressured to pursue education and to take advantage of whatever employment opportunities they found because they can no longer insure their survival by farming.

It is quite common in Sudan for migrants to travel great distances for work, from western and southern provinces to the more developed central riverain region. For Wad al Abbas villagers, labor migration is easier since they are already located in this region. The comparative proximity to the capital probably encouraged migration because of the relatively low cost of transportation and the flow of information between the city and the countryside via traders and labor migrants.

The mid-1970s marked another turning point in the development of village economy. At this time, men began migrating abroad to Saudi

Arabia, Abu Dhabi, Yemen, and Libya where oil wealth created employment opportunities. The salaries abroad are many times those paid within the Sudan. Once the first villagers went and returned with almost unheard of wealth, a steady stream of migrants began to flow in that direction. This movement is also related to wider conditions in Sudan. Wad al Abbas men are participating in a national exodus. As one villager commented on the national situation, "We used to export cotton, we used to export lots of things, now we export people."

In the fall of 1980 hardly a teacher showed up for the schools at Wad al Abbas. Rumor had it the teachers had gone to work abroad. One villager explained, "They all go to Saudi Arabia every vacation and then don't return. We used to have university-educated teachers, now we only have high school educated. The salary is so low that they couldn't support a family on it, so they go to Saudi Arabia."

Remittances are now an important part of household economy and make a difference in family circumstances. For example, in the early 1980s one family was receiving £S500 (about US$555) every three months from a son employed as a driver in Saudi Arabia. Another son earned £S1,400/month as a translator there while a third, working in Sudan as a clerk in the Kinana sugar factory, was making £S120/month. In addition to cash, migrants send clothes and luxury consumer goods such as watches, radios, and cassette players. These items, like gold, are also a form of savings for emergencies when they are sold to raise money.

In addition to the absolute decline of returns from local agriculture on the scheme, migration from Wad al Abbas appears to have been stimulated by increasing economic inequality. One young man, referring to the local merchants, commented, "if *they* left, we'd stay." An educated villager, employed in the local school, said:

> Am I a slave? Why should I work hard and get so little for it while traders and the military prosper, and others go abroad? Why should I alone sacrifice for my principles or for my country? . . . Without my brother outside [a university graduate working abroad] I would have had to migrate myself before now. Now, I want to leave by any means.

International labor migration is always temporary and the men migrate without their families. Many men go abroad for one to three years and then return to work in Sudan. Often they work abroad to raise capital in order to enter trade back home. A woman whose husband is a government employee in Sudan says, "I told him to migrate [abroad] . . . then he could become a trader in Omdurman. Wouldn't that be better?"

International labor migration from the village continues to increase and is only limited by the costs and difficulty of arranging documents and employment, and the efforts of the Sudanese and other governments to stem the tide.

The opportunities for women to earn income remain limited. It is men who have moved out of the local agricultural economy into new productive activities and relations. The majority of women are primarily occupied with unpaid domestic work for their families. Most women rarely engage in activities that yield income. This constitutes a shift in gender relations from a subsistence economy in which men and women had complementary and interconnected roles to one in which men's work is the mainstay of household economy and is, at the same time, no longer organized by the household, but through production relations outside it.

Villagers are conscious of how their lives have changed as cash has become ever more important. As an older woman put it: "Before there was duty and responsibility, now there is just studying and riches." Another woman remarked that things had improved as people became more educated, but her husband, a farmer in his sixties, contradicted her saying, "Everything has become less. Life has become less! Manners and good behavior have become less!" In another conversation, a middle-aged woman stated bitterly, "In the past, there was so much milk and butter your neighbors would give you a bowl. Now, even if you ask for it, they won't give you. People are not good like they were before." Old Haj Hussein saw local changes as related to national conditions. He said to me:

> I want money! I want to go meet the leaders in their offices. . . . I'd tell the leaders, "Don't let the lion eat the poor. You, the president, are responsible for us all." In politics, the people scream and the poor need sugar and grain and meat! Show me how a poor man like me can eat? . . . We'd go to the president, tell him, "You gave to the merchants, but they didn't let us eat. They don't leave us anything." If we went to him all together peacefully, would he be angry? We are just the poor. You go to the merchant and he tells you he's all out. Can you say anything? You go home sick and hungry and cry with your stomach aching, that's all.

While villagers retain great loyalty and pride in their community and their nation, some have become aware of the larger scheme of things in which Wad al Abbas and even Sudan are at the bottom, not the top, of a hierarchy. A young villager who trades in al Demaziin, turned to a gathering of guests at a circumcision celebration and, pointing at me said, "She comes from the biggest government in the world, you come from the smallest. Her country makes planes and such things; you grow cotton and grain and peanuts to buy planes from them." On another occasion a villager reflected:

> The English deceived us, making us farmers instead of manufacturers. They would do the industrial part. Now the machines they left us have broken and we have no spare parts because the factory that produced them stopped a long time ago. They left us in this wide world. Now the world is going forward and we are going backward.

ECONOMIC CHANGES since the 1950s have led villagers to increasing dependence on non-agricultural sources of income, labor migration, and reliance on purchased goods, including basic subsistence foods, such as sorghum. These changes were spurred by the establishment of the irrigated scheme, the expansion of the cash economy, and commercial development throughout the central riverain region, and facilitated by the growth of local educational opportunities, and, more recently, the availability of employment in nearby oil-producing nations. Today the peasants of Wad al Abbas are integrated into capitalist relations of production and exchange through the sale of their labor and products and through their consumption of purchased goods and hired agricultural labor. The village no longer has a viable subsistence economy outside the circuit of capital. The exploitative conditions imposed on village agriculture by the government scheme make villagers vulnerable to other types of exploitation by compelling them to seek off-farm income through wage-labor or self-employment in marginal commercial activities.

The major trends in the economy of Wad al Abbas since the establishment of the scheme up to the 1980s can be summarized as follows:

> *1950s.* The establishment of the irrigation scheme. The local agricultural system is completely disrupted and profit-oriented export production becomes a central feature of farming. Cotton profits and productivity on the scheme are high and villagers remain primarily involved in agriculture as the basis of the local economy. Rising standard of living, cooperative village development projects, and the growth of the

local market. Increasing reliance on purchased goods and purchased inputs to agriculture, supplied or financed by the scheme on credit.

1960s. Decline in profits and productivity on the cotton scheme followed by farmer unrest. In the late 1960s, the beginning of the emergence of a new group of wealthy merchants out of the village peasantry. High inflation and shortages of basic commodities. The nationalization of the scheme at Wad al Abbas in 1968. The start of labor migration from the village within Sudan.

1970s. Increasing labor migration and involvement in trade. Continued accumulation of mercantile wealth by a small segment. Village traders shift their businesses out of the village to towns and the local market declines in scale and regional importance. In the mid-1970s, labor migration to the oil states begins. Villagers become increasingly dependent on remittances. Greater occupational differentiation among villagers and greater disparities in wealth.

Between 1954 and 1980 the village changed from a largely self-sufficient agricultural community, loosely linked to the regional and international economy through trade, to little more than a residential unit of people linked by kin and neighborhood ties but otherwise participating fully in the national and international economic system.

Local merchants are no longer closely linked to local peasants, extending credit, purchasing produce from farmers, and selling them consumer goods. While wealthy men reside in the village or maintain wives and children there, they derive their wealth from enterprises based in the national economy. In terms of the sources of their economic power, these men are not local elites, but rather part of a national class of mercantile capitalists. Their business interests involve them in affairs outside the village and lead them to forge contacts with others like themselves. They, along with a few villagers in professional jobs, have become part of Sudan's national bourgeoisie.

The majority of the village population, however, is part of the peasant-worker class. This class is also produced and reproduced through relations of production in the national and international economic system. At Wad al Abbas its members include: farmers (part of a large national group of tenants whose conditions of production and remuneration are standardized on the many irrigated cotton schemes), petty traders, craftsmen, and assistants working in the informal sector, wage-laborers employed in

Sudan, and international migrants (part of the international proletariat working in Saudi Arabia, Yemen, Abu Dhabi, Libya, and elsewhere).

NOTES

1. Even today these are not always strictly followed if family members agree among themselves. But *shari'a* is the ultimate rule in case of dispute.
2. The *ardeb* is a measure of volume. Both a larger and smaller *ardeb* have been used at different times and places in the Sudan and Egypt (Issawi 1966). The smaller ardeb equals 12 *kayla* or 198 liters while the larger *ardeb* equals 20 *kayla* or 330 liters, using Issawi's 16.5 liter/*kayla*. At Wad al Abbas the 20-*kayla ardeb* is used.
3. Due to the difficulty of maintaining large herds on the scheme, households keep a few milk cows in the village and hire nomadic herders to graze the other animals outside the scheme.
4. Tenants on all the schemes have representative bodies recognized, and to a considerable degree sponsored, by management. At present there are two unions in the Blue Nile Schemes, the Blue Nile Tenants' Union and the Massara Tenants' Union. Wad al Abbas falls under the administration of the Blue Nile Tenants' Union.
5. While I did not witness the flogging of the Wad al Abbas boys, I observed floggings in Sudanese courts where a man wields a huge leather whip with all his might against the back of a prostrate prisoner.
6. While ownership of a tenancy is legally transferable, payment for it is illegal.
7. Tenancies may not be divided below one half, but heirs sometimes informally share produce or rotate the farming of land on a yearly basis to further subdivide a legacy.

FOUR

Off-Farm Work and Household Economy

MOST Wad al Abbas households own land and continue to organize agricultural production of their main subsistence crop, sorghum. But farming is no longer the mainstay of household economy. Household members participate in relations of production outside the household and off-farm work is essential to household reproduction. Wad al Abbas households are responsive to conditions in the regional economic system in which they participate as buyers and sellers of commodities, including labor. Moreover, households themselves have been transformed in complex ways by the process of integration into the regional and world economy, altering the social and economic basis of farming. As the household's role in organizing production has been eclipsed by capitalist relations of production outside it, gender roles, marriage and residence patterns, relations between the generations, and the division of labor within the family have been affected. Labor migration has changed de facto household size and composition, and off-farm work in general has reduced the household's agricultural labor pool. Yet the functions of the household in social and biological reproduction remain vital, and households have responded dynamically to the exigencies of off-farm work and labor migration.

The villagers of Wad al Abbas reproduce themselves and their households through the combination of household farming and the sale of labor, services, and commodities. The conditions of such production and reproduction involve household members in different types of economic relationships—as peasants producing cash and subsistence crops, as employers of agricultural labor, as workers in industry and services, and as petty traders and artisans in the informal sector. There is thus an interplay between capitalist markets for labor and commodities and household-based forms of production and exchange. As Long (1984:14) points out, "Non-capitalist forms are not, of course, outside the capitalist framework but represent the way in which local or subordinate social structures mediate the effects of capitalist penetration."

> [Households] are systems that are able to provide labor to capital precisely because they ensure the combination of income from wage labor with that from non-wage labor so as to form an adequate pool of resources guaranteeing the replenishment of labor power. . . . [T]he set of relationships that guarantee non-wage labor are themselves predicated on the world-economy. (Smith, Wallerstein, and Evers 1984:8, 11)

Although Wad al Abbas households own land and undertake agricultural production, they are not independent units of production and consumption. They purchase much of the food and other necessities they consume and they obtain cash for these purchases largely through off-farm work. Cash, moreover, is vital not only for household consumption expenses, but for agricultural production costs, especially hired labor which is required for key tasks such as harvesting.

The majority of households are unable to produce sufficient sorghum for home consumption even though it is their primary subsistence crop. Few households earn any cotton profits and, in any case, profits rarely exceeded £S40 per year in the early eighties. At that time villagers' estimates of the minimum a household could live on ranged from £S60 to £S100 per month. Household expenditures of £S200 per month were not exceptional. Oil, charcoal, spices, tea, sugar, vegetables, and meat are purchased as are clothes, tools, and furniture. Barnett reports the same to be true of Gezira households, where he found that "Almost every object consumed in the home is purchased" (1977:175). He adds, "The tenant is highly dependent on purchasing food and consumer goods from the shops because the Scheme within which he works has removed from him to a considerable degree the possibility of producing for his own subsistence and that of his family" (1977:179).

Sudanese tenant households are not unique in this respect, but rather typical of contemporary third world rural populations. Berry (1985:4) found that Yoruba cocoa farming households "are anything but self-sufficient." Many rural households in Nigeria and Ghana purchase between 50 and 60 percent of their food (Anthony et al 1979:97–98).

Throughout Africa the rural poor are often net purchasers of food (Ghai and Radwan 1983). In the Colombian region studied by Reinhardt (1988) food purchases are the major item in most peasant household budgets. In Latin America, "for the bulk of the rural poor, employment availability and wage levels are more important determinants of welfare than is agricultural productivity" (de Janvry 1981:246).

Wad al Abbas households rely on off-farm income for survival—to meet many of their consumption needs and the costs of agricultural production. As one villager declared: "There's no money [in farming]. You wouldn't find anyone just farming a tenancy. He has to have some other work."

HOUSEHOLDS AS ECONOMIC UNITS

In a discussion of agriculture in the Sudan, an ILO (1976:102) report states: "The essence of peasant agriculture is the sharing of work to be done among the members of the family: it is the farm family and not its individual members that supplies labour. . . . In short, it is not individuals who are underemployed but rather the family as an economic unit." Suliman (1975:84) similarly questions the use of the individual as an economic unit in Sudan in employment statistics, arguing that the family is the appropriate unit of analysis.

At Wad al Abbas the household is a crucial economic unit, organizing and channelling resources from production to consumption and from one economic strategy to another. The household is not, however, a bounded unit of production as peasant households are sometimes treated. The degree to which peasant households are units of production and/or consumption varies according to their participation in wage-labor and migration as well as their employment of agricultural labor. Long and Richardson describe the household as "a consumption and domestic-management unit in which labour process strategies are formulated and acted upon in accordance with shifting internal and external demands" (1978:189). They state: "As a unit of consumption and decision-making, the household is central to understanding the ways in which individuals enter into different relations of production, both within and outside the household" (1978:201).

While Wad al Abbas households are not self-sufficient units of production, they perform key economic functions. Households organize and allocate family labor to farming and off-farm work, hire and supervise paid agricultural labor, and coordinate the consumption of income and crops to meet household needs.

The great majority of Wad al Abbas households own land. A survey of a sample[1] of fifty-three households in 1982 showed that of fifty-two households whose land-holding status was known, forty-four (85%) owned land on the scheme. Eight households (15%) were landless, though some of these had had tenancies at one time. Since this study is concerned with the implications of off-farm employment for household farming, the analysis will focus primarily on land-owning households. However, as will be demonstrated in the next chapter, land is not the basis of inequality at Wad al Abbas. The economic division in the village is not between landowners and landless but between the majority of peasant-workers and the minority of merchants and professionals who are part of the national bourgeoisie. Both wealthy and poor households are among the landless and the landowners.

The household is a unit of production in agriculture in that in most cases, even where holdings on the scheme have different individual legal owners, land and labor resources within the household are pooled. Household land is generally managed as one unit in terms of decisions regarding crops, labor, and the consumption or disposition of yields.[2] The exceptions are polygynous tenants who have claims on labor in more than one household, and obligations to support more than one household from their farms; and, tenants in extended uxorilocal households because sons-in-law do not farm with their wife's father and brothers. None of the men in the sample in these affinal relationships pooled their labor or land in farming.[3] Since the multiple households of polygynous tenants share landholdings, the forty-four landed households constitute thirty-eight landholding units, which I call farming units. (In most cases, the household and the farming unit are identical and I therefore use the terms somewhat interchangeably.) Table 4.1 shows the distribution of landholdings in the sample.

Land is unequally distributed. While a tenancy is 15 *feddans* (6.3 ha), there is much variation in the size of household holdings, which range from 3 to 105 *feddans*. According to the ILO (1976) 6.3 hectare of irrigated land represents a fairly large holding by global standards. However, scheme regulations require one third of this to be under cotton each year while another third must lie fallow, leaving households with only one third of their holdings to cultivate for their own purposes.

Unpaid household labor continues to play an important part in farming at Wad al Abbas. In the division of labor by sex, women generally do not participate in household farming, although some women work as paid pickers in the cotton harvest. Village men do not pick cotton. The rest of the agricultural operations on cotton and sorghum are carried out by men and boys and, sometimes, hired workers or sharecroppers. Children's labor, though limited by school attendance, contributes significantly to farming. Households generally hire labor at peak periods such as cotton and sorghum harvests. The household coordinates this labor and the resources to pay for it, sometimes providing meals or uncooked foodstuffs to workers as well. While the scheme provides credit for cotton production, it does not lend money for sorghum production.

Cotton picking is the most labor-intensive operation in the agricultural cycle. It draws some migrants, often nomads and semi-nomads of the Butana plains to the northeast, as well as migrants from western Sudan. They supplement the labor of Wad al Abbas women and girls who are the main picking force. For other operations the household labor force of men and boys is sometimes supplemented by hired labor. Hired labor is supplied by migrant agricultural laborers, some villagers who hire themselves out, and by some settlements of Nuba, immigrants from Western Sudan, and Sudanese of West African origin[4] who, as non-natives of the region, are not eligible for tenancies in their own right. The few ex-slaves who remained after their liberation were denied tenancies, and their descendants add to the local pool of agricultural labor. Hired labor from within the village is limited as most farmers are busy on their own land. Some households sharecrop all or part of their land. Sharecroppers in-

TABLE 4.1. Distribution of Landholdings

Size of Holding	Number of Farming Units	Percent
<15f (6.3 ha)	6	15.8
15f (6.3 ha)	12	31.6
22.5f (9.45 ha)	4	10.5
30f (12.6 ha)	12	31.6
45f (18.9 ha)	1	2.6
60f (25.2 ha)	2	5.3
105f (44.1 ha)	1	2.6
	N = 38	100.0%

f = *feddan*; ha = hectare

NOTE: This includes only irrigated land at Wad al Abbas; six units own some land elsewhere.

clude both fellow farmers in the village and members of the local ethnic groups above who have no other access to land.

Economic cooperation among household members in off-farm work takes a different form than the pooling of land, labor, and produce customary in farming. There is a great deal of economic cooperation among household members, but this usually entails members contributing to common household stores from their individual productive activities rather than forming a common unit of production. Nor is household economy at Wad al Abbas necessarily synonymous with the total incomes and assets of all its members. Household members contribute to the household to various degrees. For example, a young man may not give all his earnings to his parents but keep some to save for marriage. A household member may take a limited amount of funds for household needs out of his enterprise and reinvest the rest. The economic activities of household members are potential sources of support for household needs, but the actual contribution from each source varies. In one sense, the household can be seen as an enterprise its members have in common, while having independent enterprises as well (Laslett 1984).

Though work may take household members out of the village for long periods, the household serves as a focus for their various resources. The household acts as a conduit allowing flexibility between economic strategies, channeling wealth from one sector to another and reallocating resources according to changing conditions.

The household at Wad al Abbas is predicated to some extent on the pooling of resources for consumption, the support of all members, and other functions that can generally be included in reproduction. Household labor and cash resources are allocated to alternative economic activities through the multiple occupations of individuals and through a diversity of occupations among household members. While subject to the inclinations, ambitions, and capabilities of individual household members, these activities are also tempered by the demands of household reproduction. Overall household needs, such as household consumption, means of livelihood for sons, care of dependent children, parents, or other relatives, must be met if the household is to endure and function and prepare the younger generation to become householders in their own right.

Economic cooperation and aid also occur *across* household lines at Wad al Abbas, especially between close kin. Brothers may share an interest in inherited land, rotating its cultivation among themselves or apportioning its produce, rather than dividing the property. It is common for male relatives to have business partnerships in trade and transport. In some ways the relationships between household members and with other members of their family outside the household are not markedly different. A close relation, whether sharing one's household or not, is a potential source of aid. Talking about the duty to family in the larger sense, a villager explains:

> If a man travels, he sends [money] not to his wife but to his father with instructions. The son's responsibility to his [natal] family is until death. Even married sisters must be helped if their husband is poor. Here people defend their brother and give him money even if he is lazy. They stick up for him if others criticize his idleness. Before it was even more so. We all lived and worked together. Now it is less; each person must work because development came and cooperation is less. Like *nafiir*, in the past you just provided food and drink. Now if someone is in the hospital, it is usually by money rather than by *nafiir* that his fields are cultivated.

Through their common interest in the welfare of the household, household members are linked in forms of cooperation different from other ties, however. Kinship and friendship, although infused with the norm of mutual aid, are not embodied in an institution like the household. The household is an ongoing and daily manifestation of the mutual commitment that each member shall be fed, clothed, and cared for. Thus, while in some cases cooperation across household lines is not so different from that which occurs within households—at some point each person is responsible for the welfare of his own household. To the extent that individuals do not share a household, their interests cease to be as contiguous as those of household members.

Although relations of economic cooperation, mutual aid, and support of dependents found within Wad al Abbas households are not exclusive to households, the household stands out as a primary nexus of such relationships. Furthermore, it has a physical existence—the house *(bayt)* and the courtyard *(bosh)*. It is also recognized as a social unit by villagers. In common usage various terms for family are used to refer to households, much as "family" is often used in English to mean "household." When necessary, though, speakers use terms for household, like *nass al bayt* (lit. people of the house) or *ahal al bayt* (family of the house) to make a distinction. Since "family" *(ahal)* can mean a man's parents and siblings, people often say "his children" *(awlaadu)* to refer to a man's wife and children.

The structure and layout of Wad al Abbas houses vary a great deal according to the wealth and size of the family. Usually a household inhabits a number of separate one-room structures surrounded by some type of enclosure. Indoor living space is often very small but, due to the heat, little time is spent indoors and people sleep outdoors much of the year. Some houses consist of a single square mud room with a kind of veranda in front made of low mud walls and covered by a straw roof held up on poles, called a *rakuba*. This outer domestic space is used as a cooking area and for other chores such as laundry. Such a house lacks a true courtyard *(hosh)*, so this veranda serves as women's space. Other homes have a makeshift *hosh* demarcated by a thornbrush fence. Some houses have a separate, smaller structure, often a round mud hut *(gotiya)* that serves as a kitchen. More elaborate dwellings consist of several square mud or even brick one-room structures. They might be wholly or partially enclosed in a courtyard of mud or brick walls. Closely related neighbors sometimes share a common courtyard. The most expensive housing is cement and metal. Some of the wealthiest households have multi-room cement houses with screens in the windows and large courtyards protected from view by six-foot high walls, much like urban Sudanese dwellings. This style of housing started to become more common in the late 1980s, changing the character of social space in the village by demarcating household boundaries and separating private from public space, definitively.

If a household contains more than one married couple, each has a separate room. Rooms usually stand separately as independent structures, except in the newer red brick and cement houses. Household members share a common kitchen, as well as a common *hosh*, outhouse, or *diwaan* (guest house), if any; and, they eat from common food stores. The women and children of a household generally eat together, separately from the men. Because of cooperation, hospitality, and generosity, especially between kin who are often close neighbors as well, household lines tend to blur as women of neighboring households spend much time together and in some cases eat together on a regular basis, although each woman prepares food separately at her own kitchen for this shared meal. It is also usual for neighboring women to make their *kisra* (the staple bread) in turn on the griddle *(saaj)* belonging to one of them, each bringing her own batter, and wood to add to the fire. In fact, not every household owns a *saaj* even though a woman in each household must prepare *kisra* every day.

As a result of the decline of agriculture and village-based enterprises, and the integration of village resources into the national economy, the household is emerging as a more distinct and more important unit. Villagers are now tied not only to their neighbors and kin but to employers and business contacts outside the village. The enclosed courtyards going up around some houses are one reflection of the, as yet subtle, weakening of interhousehold ties.

THE EXTENT OF OFF-FARM WORK

Male Employment and Migration

Most villagers work outside of agriculture. A small number of them are successful merchants who have established businesses. They own shops, storehouses, or trucks and are involved in long-distance and import trade in larger towns. A few men run coffee houses and eating places in markets. Many more men attempt to make a living through petty commercial activities in the informal sector. Most of these men buy and resell small amounts of goods. Some buy vegetables or other agricultural produce in Sennar and resell it in the village market. Others work in connection with more established merchants in towns, getting goods, such as cloth, on credit and peddling them in the marketplace. Other men work as shoemakers, saddle-makers, *angaraib* (traditional bed)-makers, builders, carpenters, butchers, and tailors. Almost without exception these artisanal operations are very small scale, one man working alone. The intensity of such activities varies from intermittent work to supplement farming, for example, to a full-time occupation on which the man and his household depend for a living.

Fewer men work as wage-laborers or salaried employees. Most of the workers are semi-skilled, such as drivers, while others are white collar employees including clerks, accountants, cashiers, and teachers. Much of the waged and salaried employment is in the public sector in the post office, railroads, government departments, development projects, public health services, police, and the military. There are a few professionals with advanced degrees. A small number of villagers work as unskilled manual laborers, factory workers, or hired agricultural laborers. Some of these men at all levels of employment have gone abroad to work in Saudi Arabia, Yemen, Libya, and Abu Dhabi.

Table 4.2 shows the breakdown of occupations among men in the 43 land-owning households. These households contained 93 men of whom

19 (20.4%) were unemployed and seventy-four (79.6%) were employed or self-employed at the time of the survey. The occupations of 73 of the men were recorded. A striking fact is that only about one in every ten working men is engaged full-time in agriculture. Out of 73 working men, only 8 (11%) are full-time farmers; another 12 men farm in addition to pursuing other occupations. Thus, of 20 men who do some farming, 12 (60%) have another occupation as well. This indicates, among other things, how distorted our picture of Wad al Abbas would be were we merely to consider them as a farming population, even though most households own land and produce crops.

Yet such views of rural populations are common. In the case of Sudan, Voll (1980:279), for example, states that "most of the cash in the Gezira comes directly from cotton," although she presents no data to support this assumption. In fact by 1973 it was estimated that 41 percent of the population in the Blue Nile Province (which then included the Gezira) were primarily engaged in non-agricultural activities, not even counting

TABLE 4.2. Men's Primary Occupations

Occupation	Number of Men	Percentage of Working Men	Breakdown
Trader	29	39.7	Wide range
Farmer	20	27.4	8 full time 12 part time
Wage or salaried worker	13	17.8	8 in Sudan 5 abroad
Craftsman	5	6.8	1 shoemaker 1 carpenter 1 butcher 2 builders
Other	6	8.2	2 soldiers 1 taxi owner/driver 1 bakery owner 1 private agricultural scheme licensee[a]
TOTAL	N = 73	100.0%	

NOTE: Data on men in 43 tenant households.

[a] The Sudanese government grants short-term licenses to rainland for mechanized farming for a relatively small payment.

pastoral nomads (Nigam 1977:144), while recent research on Gezira tenant households found that 70 percent of their income comes from nonagricultural sources (ILO 1984).

At Wad al Abbas, trade is the most important occupation, accounting for 29 (39.7%) of the men. Even more men actually engage in trade since many of the part-time farmers are also petty traders. Wage-labor accounts for 13 men (17.8%)—fewer men than either trade or farming when part-time farmers are included. But all of the wage-laborers are employed full time while petty traders and craftsmen are often underemployed. Two of the part-time farmers also work as agricultural day laborers.

Not only do men engage in other occupations besides farming, but these occupations often necessitate periodic absences from the village or migration. Table 4.3 gives a picture of the extent of male migration from Wad al Abbas. It shows among other things that 45.2 percent of *all* adult males in the sample landowning households are living outside the village for work or training. Thirty-four men are absent for work. They constitute 36.6 percent of all adult males and 46 percent of all employed or self-employed males.

Moreover, the figures in table 4.3 underestimate the involvement of villagers in work outside the village. There are a number of men who make a living through businesses or jobs outside the village but within frequent commuting distance. These men were not counted as migrants. The extent of migration is an indication of how linked villagers are to economic processes and relations based outside the village.

TABLE 4.3. Male Employment and Migration

Category	Number	Percentage of		
		All Men	Working Men	Migrants
Total Adult Men	93	100.0%	—	—
Employed/ self-employed	74	79.6%	100.0%	—
Migrants for work	34	36.6%	46.0%	81.0%
Unemployed	19	20.4%	—	0.0%
Migrants for school or training	8	8.6%	—	19.0%
Total migrants	42	45.2%		100.0%

NOTE: Data on men in 43 tenant households.

Women and Cash Income

Women have little opportunity to generate income. However, there are a few ways some women earn small amounts of cash. The most common is weaving *tabaqs* (woven fiber tray covers) for sale. A woman rarely produces more than 2 *tabaqs* a month, which sold for only £S5–8 (US$6–9) each in the village in 1982 and the materials cost a few pounds.[5] Some women sell roasted peanuts, fried dough, dairy products, or other foods they produce from their homes. In 1982 a woman could make up to about £S1.50/day (US$1.67) selling fried dough *(ligaymaat)*. A woman working every day, selling dairy products, could make more depending on the milk yield, but few households have milk to sell. Some women braid other women's hair for pay, earning about 75pt to £S1 (US$ .83–$1.11) for several hours work. Other women sell things from time to time such as dishes, charcoal, or vegetables. Often the goods are provided by a male relative, and the woman sells them from her home, keeping all the money. Similarly, the dairy products women sell are usually from their husband's cows. Women past child-bearing, whose mobility is not restricted by norms of seclusion, sometimes engage in petty trade, usually of *tabaqs*, in the local market. A few of these women even travel to trade in other towns.

Some older women and young girls pick cotton as paid laborers during the harvest at Wad al Abbas. The work is seasonal, arduous, and poorly paid. With the poor yields at Wad al Abbas, local pickers were making about £S1 (US$1.11) per day in 1982 for this back-breaking work. Since pickers are paid by volume, skillful pickers can earn more. However, the picking season is only three months out of the year and few women, if any, even pick for the entire season. Moreover, the majority of women who are of child-bearing age are precluded from this work by norms of seclusion and their domestic duties.

Women's petty commodity production, petty trade, and agricultural work generate little cash. They spend it on ceremonial gifts, loans to other women, and on dishes, pots, and pans which they display prominently in their homes, and use for feasts at funerals, weddings, and other gatherings.

Women also obtain cash through life-cycle celebrations. It is customary for women to give small sums to a bride, a woman who has given birth, a woman whose child has just been circumcised, the mother of a bride or groom, etc. The amounts are usually small (25–50pt in 1982), but since many women are involved, the donations add up. A woman repays these gifts over time as she attends the celebrations of others. The recipient has the advantage of receiving a lump sum. The same practices are carried out by men on these occasions and the amounts of cash involved are larger as donations are partly based on means. However, men have access to cash in many other ways. There are a number of rotating credit associations *(sanduq)* composed of women, but their operation and membership are limited by the fact that few village women have any regular source of cash.

No avenues of accumulation are open to women. At best, if they are able to come by a lump sum or save up, they buy gold jewelry which is the most widespread form of property among women. Women usually receive gold bracelets from their fathers at marriage and gold wedding rings have become part of the gifts presented to the bride by the groom. Husbands sometimes give their wives gold jewelry as well. Women value gold as a sign of status and as a store of wealth for emergencies. Few women own any other type of property such as land or livestock, beyond a few goats. There are no women of independent wealth in the village. As one woman remarked, when her son spoke of a rich woman, "a wealthy woman?—only if she has it from her family." The "wealthy" women are wives or daughters of well-to-do men who have some access to resources—cash, dairy products, food and household supplies, and gold jewelry through these men.

There are three midwives in the village, two of whom have some government training. They are paid a few pounds and also receive gifts in kind, such as soap, oil, sugar, and meat for their services. Midwifery is not considered acceptable for the average village woman, however. One of the midwives is from outside the village and was later married to a local and the other two midwives are of slave origins. One young village woman is employed intermittently as an assistant in the local kindergarten (intermittently because the kindergarten is often closed due to lack of funds).

There are also two women who earn money through their activities as *shaykha*s in the *zaar* cult of spirit possession. However, the *zaar* cult is repressed in the village since it is seen as contrary to the Islamic faith. No large ceremonies are held in Wad al Abbas. The *shaykha*s operate in their homes, treating women two days a week and receiving payment for their services. *Shaykha*s earned about £S5 for several hours work in 1982 as well as gifts and extra payments if the cures were judged effective, making this work comparatively lucrative relative to the meager earning opportunities for women. However, because of the stigma associated with *zaar* activities and the fact that recruitment is through experiencing possession states, this occupation is not an alternative for most women.

In 1981 the village had its first female high school graduates. By June 1982 all three of them were employed as teachers in the girls intermediate school at Wad al Abbas. They were the first village women to work in full-time salaried jobs. These jobs were only considered acceptable for them because they would remain in the village living with their families and work mainly with children. Even so, it was controversial and some people disapproved. At that time, no woman from Wad al Abbas had worked outside the village, although there were a few cases of village men living outside Wad al Abbas with wives from other areas who were employed.

In 1986, one of the first female high school graduates joined her husband who had migrated to Yemen and was employed along with him as a teacher there. By 1988, they had returned to the village and he was trying to establish himself as a trader in Sennar while she, expecting their third child, was involved full-time in domestic duties. Significantly, one of the other women who graduated with her also ended up working as a teacher outside the village, in the Roseires area where her husband is employed in a state veterinary station. With the increase in the education of women, their involvement in full-time wage-work is likely to become more common. In 1982 there were twenty-two girls from Wad al Abbas in Sennar high school, sixteen of whom were first-year students.

A statistical picture of women's involvement in cash-earning activities emerges from the household survey data. The female household head (wife or widow of the male household head, generally the oldest male) was interviewed in great detail about herself and asked about the activities of other household members as well. The sample of 43 land-owning households yielded a total of 72 adult women (over 16 years old). There were insufficient data on 9 of these women, leaving 63 women. Table 4.4 shows their degree of involvement in cash earning activities between 1980 and 1982. Less than one third of the women sampled attempted to

TABLE 4.4. Women's Participation in Cash-Earning Activities

Degree of Participation	Number of Women	Percentage
None	34	54
Occasional	10	16
Regular	19	30
TOTAL	N = 63	100%

NOTE: Data on women in 43 tenant households.

generate cash by their own activities in any systematic way. And, as stated earlier, the earnings of these women are small.

The growing importance of production based outside the community and the increased dependence on money in household economy has meant different things for women than for men. Women have not been able to move into trade and wage-labor and their roles in agriculture have declined as a result of the scheme (Bernal 1988a). Since the 1950s women have experienced what Rogers (1980) calls "domestication"—their contributions increasingly circumscribed to the family and household. They perform unpaid work in connection with property they do not own and commodities purchased with men's money. Elsewhere I have considered the implications of these developments for gender relations in more detail (Bernal 1985).

There is evidence, however, that a counter trend may develop as young women obtain educations that qualify them for respectable work as teachers, and villagers see the economic rewards a daughter's or wife's labor can generate. In 1988 such women were still only a handful. But the next generation of women may experience some of the differentiation already experienced by men, as a minority gain access to independent incomes while others remain limited to unpaid domestic work. While women might benefit from a reduction of their economic dependence on male relatives, ultimately the marketing of their labor will represent a further extension of external control over local resources, and facilitate a greater intensification of labor on the part of these households.

OFF-FARM WORK AND HOUSEHOLD REPRODUCTION

Involvement in off-farm work has affected household development cycles and composition. In changing circumstances, household development may be more accurately described as an open-ended process rather than a cycle (Ong 1987). The effects of participation in off-farm work on the structure and composition of Wad al Abbas households are complex as external pressures interact with local culture in the process of household formation and reproduction.

Residence patterns and choices of marriage partners are beginning to show more variation among the younger generation as some of those who work outside the village take their brides with them or marry outsiders.

The amounts of bridewealth and other exchanges associated with marriage are increasing, particularly due to international labor migrants. The age at which people marry is also rising as the cost of marriage increases and as men undergo longer periods of education and training before assuming adult productive roles.

Large families are the ideal and families of eight to ten children are usual. However, all siblings rarely reside in one household, since the older ones often marry and move out before the youngest ones are born. Fertility and the stage in the development cycle play a role in determining household size, while labor migration affects the de facto size and composition of households. The most significant effect of off-farm work on household size and composition has been the difference between the de jure household and the household members actually present in the village (the de facto household). Work outside the village has deprived the household of its most productive male members and the de facto household consists of older men, women, and children. As men are drawn into productive relations outside the household, women increasingly form the core of Wad al Abbas households and assume greater responsibility for household management.

Adult men are by far the primary income-producers in village households. Children are basically limited to domestic chores and unpaid agricultural work, with the exception of cotton picking for girls. Young boys provide important help in farming, but they work as unpaid family labor and do not hire themselves out. Boys from about age fifteen sometimes help a father or brother, minding a store or collecting fares in transport. Occasionally they are employed by a relative or fellow villager in such work, or they try their hand at petty trade. Wad al Abbas women are in a dependent status when it comes to cash. Their dependent status is highlighted by the norm that a woman's earnings, unlike a man's, should not contribute to the household, but are hers alone. The seclusion of women throughout most of their lives restricts them largely to domestic tasks which make an essential contribution to the household but do not yield income. Moreover, the demands of this work—child care, cooking, laundry, and care of the sick—on women's time often preclude consistent work on cash-generating activities.

Few women engage in any income-generating work, and the kinds of work open to women are associated with such low incomes as to be insignificant in terms of household cash needs. Many women have no income of their own. They ask their husbands for small amounts for incidentals or, if they handle any household money, keep a little aside for themselves. Men make the major household food purchases if they are home. But usually they give their wives money for daily supplies which children purchase at local neighborhood shops or in the village market.

Some labor migrants send remittances directly to their wives while others send money for their wives and children to their fathers. Men's absences give women more managerial functions in household economic matters, but often under difficult circumstances. The timing and amounts of remittances can be unpredictable and, if special needs arise, it is sometimes hard to communicate them quickly to the absent husband or son.

Although largely outside the cash sphere, the daily contributions of women and, to a lesser extent, children to household economy are extremely valuable. Women's domestic labor in cooking, cleaning, laundry, child care, and other tasks is essential in freeing men to devote their time to cash-oriented activities. Along with the production of food crops on the scheme, women's unpaid domestic work makes up what is left of the subsistence economy at Wad al Abbas.

Wad al Abbas households depend greatly on cash to fulfill their daily needs, however, and women's domestic work is predicated on household purchases from men's incomes. For example, much of the food that women cook is purchased. While crucial to the functioning of the household, women's labor does not generate wealth. It is often not recognized as work by village men, who are inclined to complain that "Women have no work. All they do is eat." Men's cash earnings determine the economic well-being of the household in a way that women's domestic work does not. And, as we will see in the next chapter, economic differentiation between households arises out of men's positions in the regional economy.

But it is the labor of women and children in maintaining households and farms that frees adult male labor for more remunerative activities. The current organization of Wad al Abbas household economies and labor migration are predicated on the fact that some members of the household do not migrate. In the case of women, social restrictions on their mobility independent of a male guardian, restrictions on their contact with unrelated males, and the limited market for their labor currently have the consequence of reserving them for this role.

The importance of women's labor to household economy is indicated by the fact that there can be virtually no household at Wad al Abbas without a woman. Yet there are households without men. Old women live alone and some widows live with their young children, often supported by male relatives in other households. However, I encountered no case of a man or men living alone in Wad al Abbas. In towns, groups of young men live *azaaba* or bachelor style, renting a place together, hiring servants, and eating meals in the market; but this is not done in the village. Within the village, a single man lives at his parents' home until marriage. A widower or divorced man remarries or a female relative forms a household with him. This is testimony to the importance of women's domestic labor.

Marriage and Residence

The economic and kinship purposes of marriage at Wad al Abbas are reflected in the customary congratulations to the couple, "*al bayt maal wal iyal*" (to the house—wealth and children). Villagers prefer marriages among kin and within the village. The ideal is for the offspring of brothers to marry (*bit amm* marriage). Marriage with other relatives, such as mother's brother's offspring, is also common. Villagers consider themselves "one family" even where the relationship is not clearly traceable, so marriage to non-relatives within the village is completely acceptable as is marriage to kin outside the village. The ideal, however, is to combine kin and village endogamy. More rarely do villagers marry non-kin from outside the village. As one young man commented, "If a person is known, his ethnicity, his origins are known, he may marry here. Otherwise they will never give [a wife] to an outsider." The emphasis on origins stems not only from the importance of coming from good, trustworthy people but also being recognized as a descendent of free Sudanese rather than slaves. One young man explained the preference for kin endogamy this way: "They [family] are easier on you. Whatever you give [as bridewealth], they accept it. They don't say it was little or it was much. They just accept it. And there are no differences. There is no one new, everyone is the same together. If you marry someone outside [the family], there may be a difference." An older villager says, "Here we believe it is better to marry your *bit amm* and your *bit khaal* [mother's brother's daughter], and better to marry from Wad al Abbas than from outside."

Some villagers have married outsiders, however. In a few cases these were outsiders who worked in the village (police, teachers, etc.). Traders whose businesses are based outside the village sometimes take wives (often second wives) in those places. With the increasing out-migration of men for work and education, this trend will probably increase as villagers are drawn socially as well as economically into the national system. Such practices are already more common in urban areas. At present in Wad al

Abbas, though, local control over marriage and the value of ties to kin and community remain strong enough so that even men who have been away for years, working or studying, generally return to select a bride from the village or from relatives nearby. In some cases they are responding to family pressure rather than exercising their own preferences. Elder male relatives virtually dictate women's marriage partners and continue to exert influence over the marriages of young men.

Residence patterns, like marriage choices, exhibit variation and have been affected by off-farm employment. The preferred and customary pattern is uxorilocal residence for at least the first few years of marriage. It is not uncommon to find a couple still in the wife's family's compound even after having had three or four children of their own. It is the responsibility of the father of the bride to provide a room and furnishings for the new couple. The kitchen and any other facilities such as outhouse, *hosh*, or *diwaan* are shared. A new bride does not assume cooking and housekeeping responsibilities for at least forty days; domestic chores are carried out by her mother and sisters. In many ways, a women remains a daughter in her mother's household after marriage.

Eventually, a husband establishes an independent household with his wife and children where he chooses. Ideally he sets up his own household near those of his father and brothers. Occasionally, a young couple moves from the bride's family household to the household of the groom's parents, if the groom wishes to move but lacks the means to establish an independent household. Other couples establish homes near the family of the bride, and some move to other areas of the village or, less commonly, outside it. Involvement in off-farm work has meant that grooms are not as dependent on the father of the bride or their own fathers to provide housing and furniture. The economic means of the groom himself have thus become a factor in residence choices, along with the relative means of the bride's and groom's families.

The growing involvement in off-farm work is altering residence patterns as some husbands relocate their families away from the village to where they work, in some cases omitting the period of uxorilocal residence altogether. However, cost alone prohibits most labor migrants from doing this since it is more expensive to maintain a household in town than in the village. Men working outside the village within Sudan can travel frequently between work and their village homes and therefore may not feel great pressure to relocate their families. For men who migrate abroad, the separation from their families and the community is much more complete. Few manage more than a yearly visit and some spend several years away before they visit or return for good. However, the conditions of their employment and migration not only make it difficult, but sometimes specifically forbid them to bring their families. Moreover, men often live in worker dormitories or share crowded residences with other men to conserve their earnings, which would not be possible if their wives accompanied them.

It is not uncommon for families to move back to the village after having lived elsewhere. People mention the high cost of living outside the village and the social distance between neighbors compared to the close ties in the village. Sometimes the move back to the village is prompted by the husband's plan to migrate abroad for work. He can leave his wife and children behind, counting on the economic and social support of kin in the village.

Uxorilocal residence, the division of labor by sex, the seclusion of women, and male labor migration mean that women form the core of the household. Women spend most of their work and leisure time at home or visiting others at home, while men's work takes them out of the house and they spend their leisure time in public places as well. Within the *hosh* women move freely, and the kitchen area is a social center where women and children gather. In contrast, the *diwaan*, where male guests are entertained, is removed from the rest of the buildings or even outside the *hosh* walls by itself.

The custom of uxorilocality appears out of place in a society that otherwise places much emphasis on ties between men. Perhaps it reflects the strong bond between mothers and daughters. The women of a household must live and cooperate very closely in domestic work—food preparation and child care. A household based on a mother and her daughters may function as a stronger, stabler unit than one where a young bride must cooperate on a daily basis with her mother- and sisters-in-law. Women are often reluctant to leave their mothers even though their husbands press for an independent household. Married women living apart from their mothers generally return to their mother's home in the last three or four months of pregnancy and remain there until at least forty days after giving birth. This is done with several successive children, so a woman spends much time in her mother's household even after marriage.

A young man and his father-in-law do not socialize and work together nor do the men of the extended family household necessarily form a productive unit as do the women. Men often pursue independent economic activities and contribute to the common budget of the household, though sons sometimes work with their fathers. Men's social activities are also much more segregated along age lines than women's. A young man spends leisure time with peers of his age group rather than with his father-in-law or father from whom he keeps a respectful distance.

Uxorilocal residence also makes sense where women are married at a very young age. Mature women in the village now were often married as children, between the ages of 11 and 14, sometimes before puberty. However, age at marriage is being pushed up slightly, to 15–17 for women, with grooms ranging from a few years to considerably older than their brides. Men's marriages are delayed by their increasing years of schooling and the rising cost of marriage, in bridewealth and gifts to the bride and her family, which takes a man time to accumulate. Since sons who work in trade or wage-labor may be better off than their fathers who are farmers, they bear growing responsibility for the cost of their own marriage. In addition, families often expect a son to defer marriage until he has contributed to improving the family's living conditions or until the next eldest son begins to earn an income.

Now that so many men migrate for work, uxorilocal residence has the advantage that during her husband's absence a young wife remains with her parents rather than being left alone with her in-laws. On the other hand, as many marriages are among kin and often close relatives (cousins), the distinction between affines and cognates is not a clear one. This is, in fact, the reason many villagers give for the preference for *bit amm* and other endogamous marriages—that all parties are relatives and therefore bound by mutual respect and obligations that transcend conjugal or affinal ties. They see this as important in preventing conflict between affines and reducing the chances of divorce.

As international migration becomes more established as a way of life for villagers, the pattern of men leaving their wives and children behind for many years is changing. In 1982 there were only one or two women who had accompanied their husbands abroad or even visited them there. By 1988 there were numerous cases of migrants who came for, or sent for, their wives and children to join them or visit them. Thus women are being drawn away from their relationships with their mothers and sisters, and the nuclear household is assuming more independence.

Bridewealth *(mahr* or *sadd al maal)* is a sine qua non of marriage at Wad al Abbas. It is always paid in cash and has been for as long as villagers remember. In the early 1980s, £S300 was considered a respectable bridewealth in the village and £S200 seemed to be the minimum. Bridewealth was rising though, largely because of international labor migrants who had the means to pay more and were paying around £S500 (US$555) in

1982. This was a national phenomenon and there were apparently attempts to discourage it through radio broadcasts. One woman said, "Didn't you hear Numeiri on the radio? He said, 'Don't marry the *mughterib* [international migrant], marry a local. The *mughterib* gives you lots of things and cheats you. He will go away for two years and not come back, but the local stays with you with the few things that he has.' " The girls listening were not convinced and found this laughable.

The economic position of successful traders and migrants may give them a broader choice of mates, as girls and their families see them as desirable husbands. The higher bridewealth paid by international migrants puts pressure on all grooms. However, the value villagers continue to place on marriage to close relatives has, thus far, softened the effects of the economic disparities between prospective grooms arising from off-farm work. Lower bridewealth is acceptable among close relatives and those with less means pool resources. Kin relationships and obligations can still outweigh purely economic considerations in match-making. Moreover, a girl's *wad amm* (father's brother's son) customarily is supposed to be asked his permission before she can be betrothed to someone else.

Villagers practice Muslim polygyny, which allows a man up to four wives at a time. However, the majority of men are monogamous and it is rare to find one with more than two wives. Polygyny is not limited to the wealthy. Among the elder generation, some poor men have second wives. Polygyny seems to be declining, however. One hears of men in the past who in a lifetime were able to have ten or more wives, but I encountered no man with even four wives. And, while among wealthy traders young men continue to take second wives, young educated men in salaried jobs appear to have turned away from polygyny. Most men cannot afford polygyny in any case, and it becomes less affordable as costs of living and costs of marriage rise.

Polygyny has important consequences for household organization as wives of the same man almost always have separate households, usually in different neighborhoods of the village, and sometimes in another village or town. This means, among other things, that polygynous men are members of more than one household. Co-wife relations are thought to be fraught with conflict and therefore it is not considered proper for a man to marry two women who are kin relations to each other.

Levirate (a woman marrying her late husband's brother) and sororate (a man marrying his late wife's sister) marriages are regarded as good solutions to the problems raised by the death of a spouse, but custom does not dictate them. A brother is not obligated to marry his brother's widow, nor is she obligated to accept him. Likewise, a widower is not obligated to marry a wife's sister, nor is her family obligated to give him another daughter in marriage.

Divorce is practiced by villagers according to Islamic law. Only in exceptional cases do they actually seek recourse to the *shari'a* court in Sennar, however. Legally, men have wide latitude in initiating divorce while women's rights to divorce are circumscribed. If the couple are uxorilocally resident, the husband returns to his parents' household at divorce, otherwise it is the divorced wife who leaves her husband's home and returns to her parents. It is generally considered that after a woman has borne several children she is not divorceable and much social pressure is brought to bear on a man who attempts to divorce such a woman. Even so, such divorces do occur. Divorce is more common and more acceptable, however, in the early years of marriage. Children remain with their mother at least until the age of seven when legal custody passes to the father. In practice children often remain living with their mothers after this age as men do not always exercise their custody rights by removing children from their mother.

By one villager's estimate, only five to ten marriages in a hundred end in divorce. Divorce may have been more common in the past since many marital histories included divorce, and it was not unheard of for a man to have divorced more than one wife. Yet between 1980 and 1982 there were only a handful of divorces in Wad al Abbas (and many marriages). Labor migration makes it fairly easy for a man to avoid and neglect a wife he no longer wants without divorcing her. Moreover, since villagers interpret Islamic law to mean that only a third divorce is final, couples are often reunited after months or even a year apart. Village and kin endogamy mean that divorce can disrupt social relationships, and wide pressures are brought on the parties to reconcile.

Men appear to have no difficulties remarrying even in cases where they were regarded as having divorced irresponsibly. Young divorcees or widows with few or no children generally remarry as well. And in this marriage, unlike a woman's first, the choice in deciding whether to accept a marriage offer is her own to make. Divorced or widowed women who have already borne several children are unlikely to remarry. Few men choose to marry such women and the women themselves profess no interest in marriage, looking instead to their sons as eventual sources of support.

Divorced or widowed women with teenage sons do not have to join their parents' household. They make up some of the female-headed households in Wad al Abbas. These also include old widows whose children have all grown up and moved out. In some cases, a granddaughter is given to them to be raised as a daughter and to help around the house and run errands. Female-headed households are among the poorest in the village.

De facto female-headed households occur as a result of husbands' migration. Many labor migrants are young men who have yet to marry, but a considerable number of married men work outside the village and leave their wives and children behind. As with divorced and widowed women, in the absence of teenage sons, the wife normally returns to her mother's household or, less commonly, stays with her husband's parents during his absence. Since most couples live uxorilocally for several years, their early children were born into the household of the wife's parents and remain especially close to these relatives.

Work abroad is particularly hard on families. The long separations it requires mean that some small children grow up not knowing their fathers. Young wives under pressure to demonstrate their fertility express concern that their husband's long absences do not give them the opportunity to conceive. Women with children struggle to be both father and mother to them. Often women in this position have special difficulty exercising their authority over sons. Batul, whose husband was working as a truck driver in Saudi Arabia in the early 1980s, complained: "Since I married, Osman hasn't spent one full month with me. I tired myself out for these kids, punishing them [like a father would] for everything wrong they do, so that they will learn to be good." As it turned out, Osman never returned to spend his old age with her and the children. In 1987 he was killed in a road accident in Saudi Arabia, leaving Batul a widow. Their eldest son now works in Saudi Arabia and helps support his mother and siblings.

Husbands and wives suffer during the separations caused by migration. A girls' folk song popular in the village tells of a woman's sadness when her beloved goes abroad: "Al musafir Jeddah, khalani baray, wai, wai, wai, khalani baray (The one who traveled to Jeddah, left me by myself [sounds of weeping], left me by myself." Nonetheless, some wives pressure their husbands to migrate because of the money. In other cases, wives and relatives do not want a man to take work far away. A newlywed wife, one of the few women employed as a teacher in the village, had this to say about migration abroad:

> We don't have such thoughts. I believe a person is better off content with what he has than striving always to get something more. Here we

are with our family. If we went away, who knows what may happen to us or to our family. Here we are together, we are *murtaah* [rested, relaxed]. I have only one room it's true. But, I don't even know how long I will live in this room. If we live in it and then there are children, God will see that we find the money to build another room for ourselves. A person shouldn't always try for big things that he is not capable of, things that require lots of money.

Work in Sudan also separates men from their families, and entails economic hardships as well. One villager, Abdel Hay, was very unhappy working in Saudi Arabia for a year, where Sudanese are sometimes insulted, called *abid* (slave), and treated as outsiders. He returned to Sudan where after six months and considerable effort he found work, earning £S50/month as a driver on a state agricultural scheme. His wife said, "Abdel Hay's expenses are £S30 per month, one pound a day for food. He often skips tea after his meal in order to save. He lives in a rented place with other people and cannot afford to buy his own bed so he sleeps in someone else's bed while two of the others share one together."

Because of the initial uxorilocal residence of most couples, male migration, and divorce, people are somewhat mobile among households which expand, contract, dissolve (as in the case of wives and children staying with her parents while the husband is a labor migrant), and reform. The patterns people follow when they do this are evidence of the strength of kin ties beyond the nuclear family household; kin relationships are more enduring than a particular co-residential arrangement.

The elder generation and particularly mature women, because unlike men they do not migrate, play an important anchoring role. Young couples, married daughters whose husbands are absent, pregnant daughters, and divorced or widowed sons and daughters join the households of their parents (or in-laws) at times. A study of international labor migration from Sudan found the role of the elder generation who remain behind caring for migrants' families to be a crucial underpinning of migration (Mahmoud 1983). Murray (1981:64) similarly observed "the sheet-anchor role of a permanently resident senior wife or widow, while the boundaries of the de facto household perpetually ebb and flow around her" in the labor-exporting villages of Lesotho.

Villagers are drawn out of the village by the conditions in agriculture that make it impossible to support a family through farming alone and by their need and desire for cash income. They are at the same time pulled back into the village by the high cost of living outside it and the difficulty of supporting a family in town where an even greater portion of household needs, particularly housing, would have to be purchased. Social ties and economic cooperation beyond the nuclear family bind people to the community and provide them with a sense of security lacking in the more transient social relationships they have outside the village. As one middle-aged man explains, "In Sennar if you make £S60/month, you rent for £S30/month. Here we live free in our own house and we have our family to help us in anything. . . . And you can have your tenancy here and grow sorghum to eat and you can keep your livestock."

In another conversation, a young man expressed similar sentiments:

> We are good people here and we like our brothers. Even girls, we like them to marry their *wad amm* so they stay together. We don't like to separate from our family. If the father has money, his sons, even if they are seven or eight, stay with him and they cooperate together to increase the wealth. But here there are no opportunities in this village for money, only outside. So a person goes [away] two months and comes back, goes for two or three months and comes back. On the *aeed* [holiday] this village is full of people. It is our way that the village you were born in, you should not leave it.

One of the wealthiest men in the village, a merchant with shops in Sennar, explains why even he doesn't move his family to town:

> I'd like to move to Sennar because the travel is tiring, but it doesn't suit us. The [house] plots there are 16 × 20 meters. It's too small and it is rare to find two together for sale. My kids are used to air here. Also another thing, any time someone [from the village] gets sick in the [Sennar] hospital, fifteen people come to see him and they say, "We won't sleep here [at the hospital], we'll go stay with so-and-so." You want your guests to be comfortable and yourself to be comfortable so you need the space. . . . Another reason is that our wives, if their sister gives birth or if there is a wedding or a circumcision, they will want to come [back to the village]. So, we might as well leave them here where they are and we'll stay in Sennar.

"Or," he adds, teasing his wife, "I could get a second wife there."

Village households have endured, assuming new forms and functions in response to changing local, national, and international economic conditions. While involvement in off-farm employment was stimulated by local agricultural decline, such employment in turn has had consequences for household farming. It is primarily through the household that the impact of villagers' involvement in new relations of production is transmitted to agriculture.

FARMING AS ONE STRATEGY AMONG MANY

Dependence on off-farm work means that the Wad al Abbas farmer is usually only one element of a household whose members engage in diverse activities and relations of production. Table 4.5 shows participation in various occupations by household. Over half the households in the sample (55.8%) have an adult member engaged in farming. However, while counted as farmers, many of these men are also part-time traders or craftsmen. Over half the households (51.2%) have at least one member engaged full-time in trade. Fifteen households (34.9%) have at least one member engaged in full-time wage work.

The importance of off-farm income is reflected in the allocation of household labor. A significant number of landowning households have no adult member engaged in farming. Table 4.6 shows the number of adult farmers per household. For these purposes, farmers were defined as adults who take some part in agricultural production beyond managerial responsibilities, on land to which their household has access either through ownership, rental, or sharecropping. Three women farmers are included; the rest are men. Any adult who engaged in farming at least part time was counted as a farmer. All these households own land, but some have

TABLE 4.5. Occupational Participation by Household

Occupation	Number of Households with at Least One Member Working in the Occupation	Percentage of Tenant Households
Farming	24[a]	55.8
Trade	22	51.2
Wage or salaried work	15[b]	34.9
Crafts	5	11.6
Other	4	9.3

N = 43 tenant households

NOTE: As these categories are not mutually exclusive, column totals exceed the sample. Polygynous men are counted in each of their households.

[a] Includes 3 women.

[b] Includes 1 woman.

TABLE 4.6. Adult Farmers per Tenant Household

Number of Farmers in Household	Number of Households	Percent
0	19	44.2
1	23	53.5
2	0	0.0
3	1	2.3
	N = 43	100.0%

no farmer; their land is cultivated by children, sharecroppers, and/or hired laborers, while adult household members pursue other occupations.

Most households have at least one member engaged in non-agricultural work, and households do *not* concentrate their labor in agriculture, but on the contrary, diversify their economic strategies. Table 4.7 shows the number of income producers (employed or self-employed) per household at the time of the survey. Women engaged in the petty trading and other activities described earlier are *not* included because their activities are sporadic and generate very little income. Three women tenants who farm and one woman employed as a teacher are counted as income producers in their respective households. The rest are men. Although the cash incomes of farmers may be little or nothing, they are included as income producers because the sorghum they produce makes a significant material contribution to the household. In both tables 4.6 and 4.7 polygynous men are counted in each of their households, since their work contributes to each of them.

TABLE 4.7. Adult Income Producers Per Tenant Household

Number of Income Producers Per Household	Number of Households	Percent
0	2	4.6
1	18	41.9
2	10	23.3
3	9	20.9
4	4	9.3
	N = 43	100.0%

NOTE: Farmers are counted as income producers here.

Only 2 households (4.6%) have no income producer, whereas 19 (44.2%) have no farmer. (The two households with no internal source of income are supported by relatives in other households.) Calculating from table 4.7, we see that over half the households, 23 (53.5%), have more than one income producer, thirteen households (30.2%) have more than two. But only one household has more than one adult farming even part-time (table 4.6). This household is a special case since it contains adult males who are deaf-mutes and therefore unable to pursue other occupations. In all other cases, additional income producers are engaged in off-farm work. Furthermore, even when there is only one income producer who is a farmer, he often has another occupation, usually petty trade or crafts.

The involvement of Wad al Abbas men in off-farm work has had consequences for agriculture. One immediate effect is the exacerbation of the agricultural labor shortage. Many households already have inadequate labor resources to meet the demands of irrigated farming which is why hired labor has played a role on the irrigated scheme from the beginning. Off-farm work can generate cash for agricultural inputs such as hired labor, but it also reduces the pool of unpaid farm labor on which the household can draw. This increases the costs of farming as additional hired labor is required.

Some individuals are able to balance off-farm work and farming. Many traders stop working or close their shops during key agricultural periods. Some get sons or other relatives to watch the shop for them while they attend to the farm. But most wage-work and work abroad is incompatible with farming. Table 4.8 shows the distribution of labor migrants across tenant households. Slightly over half the households (51.2%) have at least one labor migrant, ten households (23.3%) have more than one migrant,

TABLE 4.8. Labor Migrants per Tenant Household

Migrants per Household	Number of Households	Percent		Number of Households	Percent
0	21[a]	48.8	No migrant	21	48.8
1	12	27.9			
2	6	14.0	One or more migrants	22	51.2
3	3	7.0			
4	1	2.3			
TOTALS	43	100.0%		43	100.0%

[a] Three of these households have no male members.

and a few have three or more migrant members, indicating greatly reduced household agricultural labor pools.

The fact that landowners and their household members are engaged in off-farm work is viewed as "absenteeism" by scheme management. In 1981, an inspector in the Blue Nile Scheme office at Wad al Abbas cited "absentee tenants" as a major problem there. The issue of the labor contributions of tenant families has received considerable attention in the literature on Sudan's schemes, all of which depend on migrant labor in addition to the tenant population. The prevailing view is echoed by Kuko (1984:96) who asserts that "tenants do not want to do the work themselves and prefer to hire laborers." This view takes no account of the other productive activities in which farmers and their families are involved. The assertions that they neglect their fields or use hired labor because they prefer leisure merely distract us from the real economic conditions under which these farming households operate.

Not only is off-farm income essential to household reproduction, it plays a part in agricultural production by providing cash needed for farming, especially for hired labor. It is not a question of prefering leisure but of survival and, after that, of returns to labor. This can mean that farmers prefer their sons to work elsewhere or even attend school with the hope of improving their prospects, and are thus willing to hire labor to replace them on the farm, for example.[6]

Tables 4.5 through 4.8 do not even fully represent the diversity of household economy at Wad al Abbas. With the exception of farmers (who are always classified as farmers), individuals were grouped by their primary occupations, for simplicity. But many people engage in more than one type of activity. Furthermore, there is considerable flexibility in economic strategies over time. Men leave wage-work and enter trade and vice versa. Men work outside the village or the nation and return. Moreover, the number of income producers and the range of possible occupations vary for individual households over time as well as among households at any given time. Households contain different numbers of adults, and most importantly adult males, at different stages in their development, and men experience periods of unemployment. Thus, as Murray (1981:50) points out, "more households are directly dependent on an income from migrant earnings than exhibit the absence of a migrant labourer at any one time."

THE RESPONSES OF VILLAGERS to the conditions in agriculture, imposed by state and capitalist control, solved the crisis in household reproduction brought about by the disruption of subsistence production and the low, unreliable profits from cash crop production. Off-farm income allows households to purchase food and other necessities and even helps them to maintain agricultural production by supplying cash for hired labor. But the movement into informal sector activities and wage-labor actually was a further stage in the process whereby local systems of production and resource control were restructured to meet the demands of capitalist reproduction in the Sudan and abroad, rather than household reproduction.

The household as a unit is both strengthened and weakened by this process. The household is strengthened in the sense that it is no longer as embedded in extended kin and community socioeconomic relationships, but it is weakened to the extent that it is now subordinated to the national economic system. Having lost the ability to sustain their members through work on their own land, households in effect have lost control over their labor resources. They are compelled to allocate labor to work that is not organized by the household or community but by capital, either directly through employment or indirectly through the market.

The key function of the household is not so much organizing production as coordinating the participation of its members in relations of production outside the household, and managing the income from that work to meet household needs. Household economy at Wad al Abbas generally consists of women's unpaid domestic work, subsistence production of grain, production of cotton as a cash crop or to retain land rights, and wage-work, trade, and/or crafts as sources of cash income for the provision of household necessities and to hire agricultural labor. In the household, goods and income derived from various types of productive relations are brought together. Household farming plays a subsidiary role to off-farm work which is the primary basis of villagers' survival. While village households cannot sustain their members through farming, urban conditions make household reproduction outside the village impossible for most. Through the household, migrants are connected to their rural communities and, through the commercial and wage-labor activities of household members, rural households are linked to the regional economy.

NOTES

1. This sample is based on a statistically random sample of forty-seven household heads, some of whom are members of more than one household because they have more than one wife, which increased the sample size to fifty-three.

2. The decision maker is not necessarily the household head; it is sometimes a son who is left in charge of the farm because the father works outside of agriculture. Individuals, not households, are the legal title-holders of tenancies on the scheme. But farming is a household activity, and the legal title-holder of the tenancy does not always coincide with the actual farmer or manager of it. Sometimes the title-holder is a child. Women as well as men own tenancies, though women are very much in the minority and usually do not farm or manage their holdings themselves. An individual is limited to two tenancies, but there is no limit on household holdings. Child tenants are in part a result of this. Thirty-six (81.8%) of the 44 tenant households contain only one tenant. Six (13.6%) have two tenants and two households (4.6%) have three. Where there is only one tenant, it is usually the male household head. In five households, the additional tenant or tenants are sons of the first. In two households they are the wives of the household head. In one household a handicapped adult brother in the household is the additional tenant.

3. The unpaid labor a farmer can command thus has to do, in part, with his position in the household. In practice, some tenants (polygynous men) command labor from more than one household, while others (the co-resident affines described above) may not be able to draw on the labor of a whole household. For 29 (65.9%) out of the 44 landowning households in the sample, the farming unit coincides with the household. The remaining 15 households involve polygynous and/or uxorilocally resident men. These households form 9 farming units that do not coincide exactly with household boundaries. Of the polygynous men, four men have two households each and one man has three. The sample thus consists of 38 farming units, more than three quarters of which (29 out of 38) are households, while the other 9 are multiple households of polygynous men or subdivisions of households containing uxorilocal sons-in-law. The farming unit, as distinct from the household, is useful only for understanding the distribution and organization of agricultural land and labor in the cases of polygynous or uxorilocal landowners; it has no other existence as a unit in terms of production or consumption. Sorghum, for example, is divided among a polygynous man's households and grain stores are managed separately.

4. There are three small villages of western migrants near Wad al Abbas that were founded in the early years of the scheme: al Guntara, Korbaj, and Fungoga. Al Guntara had 23 or 24 huts in 1982, Fungoga is slightly bigger, and al Korbaj consisted of only 7 or 8 huts. The Nuba colony consists of about 30 households.

5. The dollar value of the Sudanese pound varies and declined rapidly through the 1980s with progressive devaluations. In January 1980, £S1 was officially worth US$1.25. By June 1982 it had dropped to $1.11. At the time of my revisit in January 1988, the Sudanese pound was officially worth US$.22. Unless otherwise stated, all dollar equivalencies reflect the exchange rate of the period being described.

6. This is in contrast to Duffield's (1981) report that among the Takari, sons defy their father's authority to migrate.

FIVE

Off-Farm Work and Inequality

THERE IS both poverty and plenty in Wad al Abbas. While landholdings vary, differences in access to land are not the root of inequality. Off-farm incomes and assets are what set households apart. The village itself is not a base for accumulation beyond the scale of a small shopkeeper, but prosperous men who derive their wealth primarily from outside the village continue to live in Wad al Abbas and play a role in village life. For the most part, however, the gradations of wealth among villagers represent relatively small variations within a poor, peasant-worker population, rather than class differences. These gradations, nevertheless, are significant in determining the options and living conditions of individuals and households.

ECONOMIC INEQUALITY IN WAD AL ABBAS

A minority of villagers are part of Sudan's national bourgeoisie. They are big merchants who own cars, trucks, tractors, urban property, and many head of cattle; and a small number of highly paid international labor migrants who are able to accumulate capital and property in Sudan. Through their wealth these men are able to reproduce their class positions in the next generation, primarily through providing sons with trade capital. The majority of villagers, members of the peasant-worker class, have few assets to employ sons or to pass on to them. Education and trade are possible avenues of economic mobility but require some resources to begin with and do not always lead to success.

Villagers are very aware of wealth differences within their community and these differences are readily observable to the outsider. Household variation in wealth is revealed by consumer goods such as watches, expensive clothing, television sets (run on car batteries until the village was electrified in 1988), and costly rifles. In 1981, one household even acquired a small electric generator which could light up two rooms in the evenings. With the advent of electricity in 1988, the range of luxury goods has exploded—including blenders, irons, fans, and even a few refrigerators. But, as one woman noted, "That's for people who have money. The poor will just turn on the light, that's all." Wealth is manifested through elaborate feasts when there is occasion to celebrate, such as a boy's circumcision or a wedding. The houses of the wealthy contain large cabinets displaying dishes, glasses, and cooking pots prized by village women, and used only a few times a year. In contrast, the poorest households possess only the bare necessities for daily use—a pair of *angaraib*s, a table, a teapot, and cooking utensils. Their few items of clothing hang from pegs on the walls of their mud dwellings.

Wealth is reflected in house construction and furnishings. Housing in the village ranges from traditional mud rooms with sloping walls (*jaalous*), to rooms of (unfired) "green bricks" *(toob akhdar)*, to red brick and, for only a handful of the wealthiest, cement and metal constructions surrounded by high cement *hosh* walls. In the early 1980s, a mud room cost about £S200 (US$222) to build, including labor, wooden doors and windows, and a thatched roof. A small brick room cost £S500, while a brick room with a small brick *hosh* cost £S2,000. Some of the fancier houses cost £S4,000 (US$4,440 in 1982) or more to build. Those who can afford to, bring furniture from towns, such as double beds, cupboards, and china cabinets. The bedroom set *(owdat nowm)* has become part of the furnishings provided by the father of the bride to newlyweds, among more prosperous villagers.

A few of the wealthier families have domestic servants. No villager would engage in such work for another villager, however. Servants are either from the nearby Nuba settlement or from other areas of the Blue Nile Province. A local Nuba girl could be hired for £S8 (US$10) per month in 1981.

Among men, dress is a clear indication of wealth and often occupation. Farmers wear the shorter *araagi*, usually not of fine material and often dirty; while traders, students, and others free of manual work wear long, clean *jellabiya*s, freshly ironed and of imported, synthetic cloth. Those who work the land even look physically different as the hard work in the sun wrinkles them, while the wealthy have smooth skin. Among women, there is less variation in everyday attire. For daily chores and casual visiting most wear cotton smocks and old *towb*s. Gold bracelets, one sign of wealth, are rarely removed. On formal occasions, differences of wealth are clearly reflected in women's dress; wealthier women wear fine imported *towb*s, which cost up to £S150 (US$166.50) in 1982, and sandals. (Dr. Scholl's, which sold for $40 in Sudan, were the epitome of class in the early eighties, but less than a handful of Wad al Abbas women had them.) Less fortunate women wear inexpensive *towb*s and flip-flops *(sifinja)*.

Villagers are knowledgeable about one another's economic circumstances, and new additions to a household's possessions are rapidly known throughout the community. Even children can easily name the wealthy families in the village and the owners of major property, such as cars, trucks, and large herds of cattle. A children's rhyme describes the wealth of one of the richest families in the village:

Al Hassanab, al groush turab
ma biyakulu mulaah waikab
biyakulu rariif bi kebab.
al genaih fil dowlaab
al fika bi warra al baab.

The Hassan clan, money like sand
They don't eat porridge and okra stew
They eat good bread and meat, too.
In the cupboard, their bills are stored
They keep the coins behind the door.

It is telling that food is mentioned in this rhyme because many households get by on meager, monotonous diets and children get the less desirable food, coming after men and then women in the hierarchy of food distribution. Bread, at 5p for a small loaf, was a luxury many families could not afford. As one mother remarked, "With so many kids, who can afford to feed them bread?" In common parlance *shab'aan* (sated, full) is used to mean "rich," implying that the most basic difference between haves and have-nots is the former's freedom from hunger.

Villagers associate wealth with occupation and they equate farming with poverty. A thirteen-year-old boy summed it up as follows: "[In this village] there are teachers, government employees, traders, *mughteribiin* [international migrants], and farmers. And the worst work is farming." Another man, when asked about his siblings, replied: "All of them are farmers; all of them are poor." A college student who often visits relatives in the village reflected: "Some people work all year in the fields, hardly relaxing or enjoying anything. And, at the end of the year when they go to collect their [cotton] money, maybe they're told they're in debt. They can't even get food. While others drive Mercedes[1] and have all kinds of enjoyment and entertainment." Hard work is related to poverty; the poor are referred to as tired, exhausted *(ta'baaniin)*, while the well-to-do are relaxed, rested *(murtaahiin)*.

Kinship ties and neighborly obligations soften the impact of wealth differences to some degree. Those who have more give more in social situations involving gifts or hospitality. The well-to-do make generous cash donations at life-cycle celebrations. Donations from rich and poor at one 1981 wedding came to nearly £S3,000. On the Islamic holidays, wealthier households send a share of the meat of the sacrificed sheep to relatives or neighbors who can't afford to slaughter. Some villagers practice *zakaat*, Islamic tithing, where a portion of grain or other wealth is given to the poor at the end of the year, usually to relatives and neighbors. As one farmer explained, "When you thresh your grain if it comes to more than 40 *kayla*, you give one for every 10 to the poor. . . . It is not just to any poor person, but to someone who is linked to you in some way. You give it to their child or send one of your children with it." He added that he had sent one or two *kayla* to people that year. He himself, however, was a poor farmer. When asked why he gave to the poor rather than having them work for him, he said. "He probably has his own work, his own tenancy or something, but he didn't work well or it didn't turn out well."

Some well-to-do sons of the village who have settled permanently in Omdurman with their wives and children pool their *zakaat* and send it to a village relative with a list of recipients and their respective shares. In 1981 they sent £S5,000. International labor migrants also spread largesse beyond the boundaries of their households. As one villager remarked, "Those who go to Saudi Arabia, they know [what is required of a good person]. They don't let their children of their relatives or even their neighbors go without or go hungry."

Nonetheless, when rumors of corruption circulate and essential consumer goods are sold at black market prices or hoarded, villagers remark ruefully, "The big eat the small," and "The strong take from the weak."

Economic Mobility

One response of villagers to their growing dependence on off-farm income and wage employment is a greater effort to educate their children. A successful merchant says, "Now even a poor person will work to let their son get educated by any means possible. Before, they took him out to help with the farm or the shop. The one who is ignorant holds on to money, the ones who are educated, give out money [in order to get more back later]." Actually, education is not particularly important as a means of maintaining wealth in the most prosperous families; sons with little education are able to create lucrative businesses with family trade capital and connections. For those lacking such capital but with some income beyond subsistence, the education of children, qualifying them for government jobs in Sudan and skilled or professional work abroad, is an important economic strategy. On the other hand, especially when circumstances are pressing, young boys want to leave school and start work as soon as possible. A mother, upset that her 16-year-old son wanted to drop out of intermediate school to go to work, says when she asked him why, he answered, "to put flesh on my bones." A few months later her 14-year-old son said he wanted to leave school. She tried to discourage him, explaining, "A person who has trade, who has a car [for transport business], leaves school—no harm; but he has nothing."

Another teenage boy whose father, a small trader and farmer, had recently died in an accident said, "It doesn't seem I should study any more, I'll open [my father's] shop, later I can go to school." A migrant home on vacation from Libya explained that he had left legal studies at the University of Khartoum when he got a chance to work abroad, adding "We're not like over there [the U.S.], if someone finds an opportunity he leaves high school and even intermediate school for work."

Other avenues of mobility include trading on credit, labor migration to raise capital for trading, and apprenticeships, whereby a young man works for a wealthy trader or transporter, learning the business and making connections while building up some capital, often through getting a share in the profits of his employer. One man describes his brother Ahmed's work history as follows: "When he was small he used to work with his father in the fields. Then he said 'I don't want to.' and when he was eleven or so, he began to work as an assistant with Mohammed al Khalifa [a wealthy trader of Wad al Abbas] fixing the lorry and riding with him. After some years he got a license for small cars and after two more years he got a license for all vehicles." In 1981, Ahmed was working as a driver and thinking about trying to get work in Libya. He already had managed to get an international driver's license and a Sudanese passport.

The work histories of many village men do not progress in such linear fashion — their course is diverted and interrupted by chance reversals of personal fortune, by the hardships of migration and the lack of work at home, and by fluctuating national economic conditions. Babikir's case is illustrative. In the early eighties, Babikir was the sole support of his family of nine children; his eldest son was still in high school, and they were struggling. Babikir had not received a tenancy when the scheme was established because he was away and, according to his wife, Khadija, no one stood up for his land. They occasionally rent a tenancy from other villagers. As Khadija explains, Babikir has had a hard time trying to make a decent living:

> We went to Wad Medani once and stayed four months. Babikir's brother was living there with his wife, and Babikir drove one of [his brother's] in-laws' taxis. But Babikir is a *mowlana* [religious man] and doesn't like lots of women so he said he didn't want to drive taxi. His [other] brother called us to join him for his marriage to a relative in the north, and then we came straight back here, not to Medani. Babikir's brother began to drive the taxi himself, and eventually bought it. . . . Babikir was a driver here for al Khalifa and Abdel Karim Sherif [Wad al Abbas merchants], and he brought any new lorries down here from Port Sudan [where imported vehicles land]. Then he went to Saudi Arabia for 7 months. He didn't have a visa or a contract, he just went on the *umra*[2] and came back with £S300. In those days [1977], that was a lot of money. He asked [my brother] and some others to help him and he bought a lorry for £S800, then he worked [as a transporter], paid them back, and it became his. He sold it last year [1979] for £S2,200. I told him to become a trader but he refused saying, "I don't know trade and I didn't study." Then, he decided to buy a small car, thinking the work would be comfortable. . . . He bought the taxi and had it repaired, and even painted, in Medani for £S900. The other money was just used up [for household expenses]. We didn't know the taxi would only give us basic expenses, nothing to build a *bosh* or anything with. Now, he wants to sell it and go to Saudi Arabia on the *umra* to see if he can work.

In 1981, Babikir did sell the taxi and go to Saudi Arabia for the second time. But months passed without his finding work and he was able to send very little back to his family. He returned home after less than a year without much savings, and began to look for work locally again as a driver, cultivating a rented tenancy with his young sons. Thus, even migration abroad—perhaps the best opportunity open to a man—is not guaranteed to improve his lot. Khadija says, "Maybe if he had stuck with one thing, he'd be well-off now. If he can just find work driving [a truck] for a trader, we'll be o.k." When I met the family again, in 1988, Babikir was working locally as a driver and cultivating a tenancy he had purchased in 1985.

The vagaries of the labor market are such that, unlike Ahmed and Babikir, many villagers not only move from job to job, but from one type of work to another. Villagers are occupationally and geographically mobile; however, their economic movement is for the most part lateral rather than upward.

Since women have little chance to possess or accumulate wealth, the economic circumstances and class positions of households rest on men. Whereas education represents an investment in a son's future, a daughter's well-being is primarily secured through marriage. Some intermarriage among the wealthiest families in the village has served to concentrate their wealth and economic distinction from the rest of the village. Quite recently the education of daughters has been pursued by at least one of the very wealthy families and a handful of moderately comfortable families that had already been educating their sons. It is significant that educating daughters to the secondary school level was pioneered not by the wealthiest villagers but by those of moderate incomes. The poorest simply cannot afford school fees, dormitory expenses, tutoring, and other costs of education. The education of daughters suggests that some families are responding to economic pressures by seeking to convert more of their labor to cash. A time is approaching when withholding the labor of women from the market will be a luxury villagers cannot afford.

By 1988, Babikir's daughter had graduated from high school, was teaching in the village school system, and contributing money to the family. Several other women had graduated as well and they, like the first women graduates some years before, became teachers in the village schools. For a wealthy family, the monthly salary of £S188 (US$41.36; less than $2/day) in 1988 would be insignificant, but for other families, this added source of income could increase their economic security. Babikir used some of his daughter's earnings to build a brick addition to a room. People say even a daughter who uses the money to buy her own clothes and soap eases the burden on her family.

The material standard of living at Wad al Abbas is rising, mainly because of the growing number of men working abroad. Nevertheless, few villagers have permanently escaped a peasant-worker existence. The economic disparities among villagers, moreover, are escalating.

Off-Farm Resources

Economic inequality at Wad al Abbas stems primarily from off-farm sources. One evidence of this is the unequal distribution of property other than local farmland. Table 5.1 shows the major kinds of property owned by villagers and its distribution among 43 tenant households. Livestock is the most common form of property. In most cases, however, the animals merely contribute to household subsistence through dairy products, rather than yield income. Business properties range from stalls and shacks to well-constructed stores in towns. Even structures with little resale value spare the trader from paying rent. Vehicles are a valuable money-making asset since traders, consumers, and passengers all over the country depend largely on private transporters. Table 5.1 does not fully reflect inequalities in the distribution of property, however, since some

TABLE 5.1. Ownership of Property Other than Local Land

Type of Property	Households Owning	
	Number	Percent
Livestock[a]	23	54
Business property, store, warehouse, office[b]	15	35
Private agricultural land[c]	6	14
Vehicle: car, bus, truck, tractor[d]	5	12
Scheme land on another scheme[e]	3	7
Residential property other than family residence	3	7

N = 43 households; column totals exceed this because some households own more than one type of property.

NOTE: Property of polygynous household heads is considered jointly held by households and is thus overrepresented here.

[a] About half of these own only a couple of animals. 16 (37%) own more than 2 cows or 10 goats.

[b] 3 of the 15 own more than one.

[c] *juruf* and *jazira* land that is no longer cultivable is not included.

[d] 1 owns more than one.

[e] on neighboring Suki Scheme.

households own more than one type of property. Moreover, differences in household income and capital are not always reflected in property ownership.

To examine wealth differences systematically, a comparative measure is required. To construct such a measure each household was scored on a scale of 1 to 4 from poorest to wealthiest, based on a qualitative evaluation of their apparent wealth and standard of living observed during one or more visits to the household. Each household was also scored from 1 (low) to 4 (high) on their ownership of property, apart from scheme land.[3] The addition of these two scores yielded an ordinal scale of 2–8, with 2 representing the poorest household and 8 the wealthiest. Tenancy ownership was purposely *excluded* from this measure so that household wealth and land ownership could be considered independently. The qualitative assessment of wealth was necessary because property ownership does not reveal liquid capital or the value of commercial inventory (the key resources of traders); nor does it always reflect the salaries of *mughteribs*. Moreover, because of the importance of trade capital as a productive asset, there are well-to-do households that own no property. For example, a number of successful traders have no stores or trucks of their own but act as middlemen organizing large shipments of goods from one market to traders in other markets.

It was not possible to obtain reliable data on household income for several reasons. Villagers were sometimes reluctant to discuss their incomes in detail, partly because interviews were rarely completely private, and to discuss their resources would have exposed them to claims upon them. As one local saying goes, *al hagak, fi butonak*, "What is yours is what is in your belly." For this reason household heads may not always know what a son's or son-in-law's earnings are. Conversely, those of little means do not want to explicitly reveal their dire straits in front of others. Many villagers also say they keep no account, even in their heads, and merely get money in various amounts at irregular intervals, and spend it as they need it. Since the remittances sent by migrants often vary in amount and timing, and the earnings of petty traders and craftsmen fluctuate, it makes sense that people may not think in terms of fixed expenses or income, but simply spend more when they have it and less when they don't. As much commercial activity involves the black market as well as official markets and often entails circumventing various forms of red tape, restrictions, and duty fees, as well as unreported income for tax purposes, the reluctance of individuals involved in trade to give detailed economic data is understandable. Farmers and wage-workers are much less hesitant to report their earnings.

The differences in wealth among village households are for the most part small variations within the peasant-worker class. Only the wealthiest households, with scores of 7 and 8, are members of the bourgeoisie or petit bourgeoisie. Differences of one unit, between a 2 and a 3 or between a 5 and a 6, are not very great, while a difference of several points is substantial. This measure of wealth understates wealth differences since a household with a wealth score of 8 is more than four times wealthier than a household with the lowest score, 2. The difference between a 2 and an 8 is the difference between poverty and plenty.

Table 5.2 shows the distribution of households by wealth, using these scores. Most village households are poor, and the well-to-do constitute a small minority. Over 60 percent of the households are concentrated in the two lowest wealth categories while 14 percent are in the top two.

If wealth differences at Wad al Abbas are primarily based outside of agriculture, we would expect the 15 poorest households with scores of 2 (table 5.2), to be highly dependent on agriculture. At the other end of the spectrum, we would expect none of the wealthiest households to be primarily engaged in agriculture. Table 5.3 shows the distribution of occupations among members of the fifteen poorest households and the nine wealthiest households in the sample. Farming is the main occupation in the poorest households, while trading is the preeminent occupation in the wealthiest households. Seventy-nine percent of the income earners in the wealthiest households are engaged in trade, while only 13.6 percent of those in the poorest households are. Moreover, while all of these households own land, only 10.5 percent of the employed and self-employed individuals in the wealthiest households are farmers. In contrast, 59 percent of the economically active in the poorest households are farm-

TABLE 5.2. Distribution of Households by Wealth

Wealth Score	Number of Households	Percent
2	15	35.0
3	11	25.6
4	3	6.9
5	5	11.6
6	3	6.9
7	2	4.7
8	4	9.3
	N = 43	100.0

2 = poorest; 8 = wealthiest.

TABLE 5.3. Occupations Among Poorest and Wealthiest Households as Respective Groups

Occupation	Employed or Self-Employed in:			
	15 Poorest Households as a Group		9 Wealthiest Households as a Group	
	Number of Individuals	Percent	Number of Individuals	Percent
Farming[a]	13	59.0	2	10.5
Petty trade	3	13.6	0	0.0
Trade	0	0.0	15	79.0
Wage and salaried work	4	18.2	2	10.5
Builder	1	4.6	0	0.0
Licensee[b]	1	4.6	0	0.0
Total	22	100.0%	19	100.0%

NOTE: Data from 24 tenant households. Poorest households all had wealth scores of 2; wealthiest households' scores ranged from 6 to 8.

[a] Four of these men are also petty traders and one man is a part-time craftsman.
[b] This man organizes commercial rainfed agriculture on government property.

ers. These data strongly support the association of wealth with trade, and poverty with farming. They indicate that household wealth at Wad al Abbas is largely the result of non-agricultural activities and resources.

The following case studies of four households reveal the economic variation among households and the economic diversity within households. The wealth scores of these households are 2, 2, 5, and 8, respectively. The gap between the wealthiest household and the other three is qualitative, not just quantitative. The wealthy family is part of the bourgeoisie while the other three are peasant-workers.

MUSTAFA AND AMNA—
A POOR FARMING FAMILY

Mustafa and Amna's household is a typical poor farmer's household. They have little off-farm income. Their household includes Mustafa and Amna (husband and wife) and their five sons. Amna gave birth to girls as well but they did not survive. Both husband and wife were born in Wad al Abbas. Amna is her husband's *bit khaal* (mother's brother's daughter). Their house, a traditional mud room without a *hosh*, is next door to Mustafa's two married sisters and their husbands who are not relatives. Mustafa, about forty-three years old, received three years of primary education as well as seven years of *khalwa*, and says he can read and write. Amna is about forty years old and has no education. Their five sons are all in school.

Mustafa is the sole support of his household as his sons are too young to work. He is a farmer and a petty trader. He travels to Sennar, Khartoum, Port Sudan, and even Wadi Halfa to buy goods and sell them; but his trading activities provide barely enough cash for his family to get by. Mustafa cultivates one 15 *feddan* (6.3 ha) tenancy, half of which is registered in his own name and the other half inherited from his father. Mustafa once owned a full tenancy in his own name, but half of it was taken away by scheme management in 1981. The official reason for this was negligence in cotton production, but Mustafa says higher cotton yields are unattainable because the irrigation provided by the scheme is insufficient.

Mustafa hires workers from outside the village to help with his required 5 *feddans* (2.1 ha) of cotton, but cultivates his sorghum alone. Not counting picking, he spent £S50 (US$62.50) in 1980–81 and £S40 (US$44.40) in 1981–82 on hired labor for cotton production. However, cash credit from the scheme covered these costs. In 1981, his cotton yield was only 5 *kantar;* he says the scheme's pesticide spraying spoiled the crop. That year his sorghum crop was devastated by birds and he harvested nothing. The next year, his cotton again yielded only 5 *kantar.* "It's not enough," he says. "I'll be in debt. The pesticide is expensive, and the credit. Each tenancy gets £S600 . . . for water and everything. Five *kantar* make £S510, so they want money. I owe £S90. No profit at all, only exhaustion *(ta'ab)*." The £S90 brings his accumulated debt to £S100 or £S150—he's not sure.

His 5 *feddans* of sorghum yielded only four sacks in 1981–82, not enough to feed the family. "The water didn't come," he says, to explain his poor grain yield. They buy the rest of the sorghum they consume. Food is their major expense, and they spend about £S75 each month on food alone. Mustafa keeps the money for household expenses unless he is leaving the village, then he gives it to his wife, Amna.

Amna has no source of income of her own. She does not pick cotton or take any part in farming. Neither she nor Mustafa own any livestock or other property besides the tenancy. I ask Mustafa if he has any savings and he replies, "Only what's in my pocket, and in my pocket is nothing." This family lives from hand to mouth.

As Mustafa's young sons mature, they will be able to help him with the farming, and eventually, to earn income. But given Mustafa's limited means, his sons are unlikely to receive sufficient education or opportunity to improve the family's circumstances significantly.

HASSAN AND FATMA — A PETTY TRADER'S FAMILY

Like many other Wad al Abbas households, Hassan's ekes out a living from various sources. Its members engage in farming, trade, wage-labor, and petty commodity production. They are slightly better off than Mustafa's family. The household consists of Hassan, his wife Fatma, their three unmarried children (two sons and a daughter), and their married daughter, Muna, and her husband. (Hassan and Fatma have another married daughter who has her own house, near her in-laws. Two other daughters died, and they also lost a son.) Their two sons are in school, the smallest daughter has not started school yet. Muna is pregnant with her first child. The family lives in two mud rooms, one for each married couple. The kitchen is a *rakuba* (thatched roof extension on a room) and there is no *hosh*.

Hassan and Fatma are the offspring of twin brothers (a *bit amm* marriage, ideally preferred in the village). Muna's husband is her *wad khaalat* (mother's sister's son), so hers is also a marriage of cousins. Two of Hassan's brothers and his sister live nearby. (Only one of his siblings lives outside the neighborhood—a brother who divorced a wife here, married a woman in another village, and settled there.) Fatma's siblings, two sisters and a brother, also have their homes close by.

Hassan and Fatma are both in their late fifties. Neither has any education; they are illiterate. Hassan, a farmer and a petty trader, is the main support of the household. Muna's husband, a head clerk in a government ministry in Khartoum, earns about £S65 a month. He is away most of the time, but sends Muna some spending money now and then. Hassan mainly trades in Sennar, commuting daily from the village, except Fridays (the Islamic Sabbath). He buys several types of beans and dried okra from farmers, and sells them wholesale to traders. (He does not buy from fellow villagers because they do not produce these crops.) Hassan works alone. Lacking a shop or other facilities, Hassan is the kind of trader called a *waggaafi* (lit., standing); he works on his feet. Because Hassan is striving to build up trading capital, he spends only the bare minimum needed for household expenses, and reinvests the rest of his earnings in trade goods. He says household expenses run about £S5 a day (£S150/month), mainly for food. Hassan says he makes as much as £S10 or £S15 in trade on an average day, although I suspect he only makes this much profit on a very good day. Once in a while Hassan goes to Port Sudan to buy fuel to sell.

In the last few years Hassan has been getting more involved in trade and moving away from farming. He owns one tenancy (15 *feddans*/6.3 ha) and shares another tenancy that he and his brothers inherited from their father and farm by yearly turns. From 1972 to 1980, Hassan also cultivated additional land as a sharecropper on another villager's tenancy. But in 1981 he stopped sharecropping. "In fact, I hardly farm the land I own any more," he says. As a matter of fact, Hassan planted no cotton in 1981 and 1982. When I asked how he could manage (since farmers who fail to produce cotton can lose their land), he replied that he knows how to handle scheme administrators. Actually, he is involved with the administration, since he is a member of the scheme's Cotton Production Board (*Lajnat al Intaj*). Through these connections Hassan has been able to ensure that his land will not be taken away. As it turns out, he didn't even cultivate the other 5 *feddans* of his land fully in 1981 and 1982. He planted only half of it with sorghum each year. Trading takes up most of his time, and hired farm labor is expensive, he explains.

Hassan's two young sons do most of the farming with occasional help from him. Hassan hired no labor in 1981 or 1982, which limited what he was able to achieve. "I'm too busy to farm and when it's just my sons alone, they aren't capable," he says. In 1980–81, his sons planted 2.5 *feddans* (1.05 ha) sorghum but the crop failed, so they planted again, this time with coriander (as a cash crop). It was already too late in the season, however, and that crop failed, too. Hassan purchased all the sorghum they consumed that year. The next year, 1981–82, his sons again sowed 2.5 *feddans* sorghum. If Hassan had been able to help with the planting, they could have sowed all 5 *feddans*, he says, but he was trading in Port Sudan at planting time. The 2.5 *feddans* yielded 8 sacks of sorghum, which they kept to eat and supplemented with another 5 sacks they purchased.

Neither Hassan's wife, Fatma, nor his daughter, Muna, take part in either family farmwork or the cotton harvest. But when she can, Fatma makes *ligaymat* (fried dough balls) and sells them from the house. This nets 45p (US$.50) profit for a morning's work. Muna has no source of cash other than the money her husband sends her from time to time. No one in the household owns any livestock.

This household is making small gains, but the odds of Hassan ever becoming more than a petty trader are slim. He will not be able to set his sons up with trade capital, and they will have to find their own means of making a living.

OSMAN AND KATIIRA—
A FAMILY WITH WORKING SONS

The household of Osman and Katiira is representative of villagers with modest off-farm incomes. Their household comprises themselves, their five sons, three of their daughters, and Osman's mother. (A fourth daughter is married and lives with her husband near his relatives elsewhere in the village.) The family dwelling consists of one red brick room, one traditional mud room, and a mud kitchen enclosed by a brick *hosh* wall. Katiira's mother lives next door.

Osman's mother and Katiira's mother, both from Wad al Abbas, are sisters; thus Katiira is Osman's *bit khaalat*. Both Osman's and Katiira's fathers were born elsewhere and married into the village, something rather unusual. Osman's father was a trader from the Northern Province who came to Wad al Abbas to buy grain, which he sold in Omdurman and parts north. He had two other wives in the north. Katiira's father came from the Gezira.

Osman is about sixty years old. He says he is able to read and write although he received only *khalwa* education. Katiira is around fifty and received no education. Two of their sons and two of their daughters are in elementary school. The third daughter, now sixteen, dropped out of school after only a year. One young son completed elementary school but went no further. The eldest two sons, Ahmed and Al Tayeb, are in their early twenties; each has three years of schooling. Al Tayeb has been working in Jordan for the past two years, as a mechanic. He acquired his skills on the job in Khartoum where he worked for a while before migrating. Ahmed works with his father in Omdurman where they sell charcoal.

Osman and Ahmed rent a place to sleep in Omdurman as they spend much of their time there, but Osman comes home to his family every week. When asked how much he earns, Osman says, "We don't keep accounts, only in a rough way, but it adds up to £S500 or so in a month." Ahmed gets no share of the money they make, but Osman says, "for marriage, I will give him." The other son, Al Tayeb, sends money from Jordan now and then. Recently, he sent £S1,000 (US$1,110) which, Katiira says, they "only used for food and such." In the village, the family spends about £S7 or 8 per day on food, Osman estimates, (£S210–240/month). When he leaves the village, he gives Katiira the household money and she sends the children to make the necessary purchases.

Katiira does not earn any money herself. She says she is ill and cannot even manage household chores, "I do nothing. I lie on the *angaraib*, day and night. My daughter cooks." In 1980–81 two of her daughters picked cotton, earning about £S15 each which they gave to her; but the next year they didn't take part in the harvest.

Osman owns a 15 *feddan* (6.3 ha) tenancy. He says another 15 *feddans* were taken away by the scheme some years back. His young sons do the farming for him. In 1980–81, his 5 *feddans* of cotton yielded only 3 *kantar*. He spent £S50 on hired labor, not counting picking. His sons planted only 2.5 *feddans* of sorghum that year and they worked it alone. Birds got to the grain, however, and they harvested nothing. Osman had to purchase all the sorghum the household needed. He bought the grain in Sennar since prices are usually lower there than in the village market. In 1981–82, his sons again required some help to cultivate cotton. Osman hired migrant laborers as well as workers from a nearby settlement of Western Sudanese. His labor costs for 5 *feddans* of cotton came to £S75, not counting picking, and the yield was only 5 *kantar*. His sons once again did all the work on 2.5 *feddans* of sorghum, and it yielded 7.5 sacks. The family will eat all of this and will have to buy several sacks more.

When asked about memberships in any local village or government organizations, Osman says, "No, and I don't want to [be a member]. It's the work of thieves, nothing more. I refused it all."

The household owns a few sheep and goats, and probably several cows as well, since they have a calf at home. (They may choose not to mention their cows because livestock are not welcome on the scheme.)

The household of Osman and Katiira includes many dependents, but there are three income-producers to help carry the burden. If Ahmed and Al Tayeb remain unmarried and continue to contribute to their parents' household until their younger brothers are able to work, the family's situation will remain relatively secure. By that time, Osman will probably retire from commuting and trading and take over farming his land himself. A daughter's marriage would also bring a son-in-law into the household as another source of support.

ABDELSALAAM AND HAWA — A WEALTHY MERCHANT'S FAMILY

Abdelsalaam and Hawa's household—one of the wealthiest in the village—includes all of their ten children, seven daughters and three sons. One daughter, Awadiya, is married and lives there with her husband, Hussein, and their two small sons. Abdelsalaam's is a *bit amm* (patrilateral cousin) marriage. His daughter's marriage is exogamous, however; Hussein is unrelated to them and comes from a neighboring village. A second daughter recently became engaged to a neighbor's son, also no relation. Abdelsalaam expects his future son-in-law to take up uxorilocal residence and plans to have a dwelling built for the couple.

The household boasts a large brick *diwaan* as well as two brick rooms, with a screened-in veranda for the parents and unmarried children, plus a third brick room with a small brick *hosh* where the young married couple sleep. The shared kitchen is a more modest structure but also constructed of brick. Abdelsalaam's maternal relatives, along with many of his siblings and their spouses, live in the neighborhood.

Abdelsalaam, a merchant in his late forties, can read and write. He has three years of *khalwa* education. Hawa, about forty, has no education and is illiterate. Their married daughter, Awadiya, is nineteen, and her husband, Hussein, is in his early twenties. Both are literate; she has five years of schooling, and he is a high school graduate. One son, Jamal, has a junior high school *(senawi aam)* education. The other sons are still in school.

Abdelsalaam owns two stores in Sennar and one in the village, mainly used as a warehouse. One of the Sennar stores carries cloth, the other one soap and household supplies. Both businesses sell wholesale *(jumla)* to traders. Jamal works with his father in Sennar. "Not for a salary, but as a partner. With my son, it [wealth or profits] is all one," Abdelsalaam explains. "If he wants to have his own money separate, I'll give him, but if he needs money for marriage or something, I'll take care of it for him." Thus, in effect, Abdelsalaam controls the money he and his son make. Hussein, the son-in-law, is also a trader and has his own fabric store in Sennar; Abdelsalaam provides him with capital and they split the profits fifty-fifty.

Abdelsalaam gets his goods from Khartoum, Wad Medani, and Port Sudan. Sometimes he goes to get the goods himself; sometimes he places orders by phone to companies and pays by check. He has two bank accounts. One is in his name alone, and another is a joint account with his son-in-law.[4] He does not save. "My money is always working," he says. He had a short-term credit line up to £S20,000 (US$22,200) in 1982.

In addition to his Sennar stores, Abdelsalaam buys sorghum from traders in the *saiid* (upstream, further south), where it is grown on private rainfed mechanized schemes, and sells it here in the village.

Abdelsalaam employs several workers to help in his businesses. His mother's sister's son *(wad khaalat)* is the overseer of his warehouse in the village, and receives 2 to 5 percent of the profits. (This could not amount to much, but the job is only part time and the overseer has other sources of income.) Abdelsalaam also has two hired men working in the Sennar shops. One man is from Wad al Abbas, the other is from Sennar, "distant relatives," he says.

Abdelsalaam owns thirty head of cattle. He keeps them in the village and employs two hired herders to tend them. As is usual at Wad al Abbas, the hired herders are Beni Amer nomads from eastern Sudan. The cattle provide milk for household consumption and income from the sale of old cows and male calves. Each cow costs about £S50/year for herding and for hay to supplement grazing in order to keep them fat and producing milk—Abdelsalaam thus spends about £S1,500 per year (US$1,665 in 1982) just to maintain his cows.

In 1982 Abdelsalaam was considering starting a private rainfed mechanized scheme to grow sorghum and sesame on about 500 *feddans* (210 ha) in the Lakendi area further south, where rainfed agriculture is possible. Such schemes are always worked exclusively by paid laborers. As it later turned out, this plan was never implemented.

Here at Wal al Abbas he owns one 15 *feddan* (6.3 ha) tenancy which he inherited from his father, as did each of his brothers. In addition, he bought (illegally, since tenancies cannot be transferred for money) half a tenancy, 7.5 *feddans* (3.15 ha), in his son's name for £S120 in 1980. It must be good land and well-positioned in the irrigation system since an entire tenancy generally could be purchased for £S100 at that time. He and his children do no farming at all. They hire itinerant workers who come to Sennar to cultivate the land. In 1980–81 their 7.5 *feddans* (3.15 ha) of cotton yielded only 5 *kantar* although he spent over £S100 (US$125) for labor, which exceeds the cash credit provided by the scheme. He received no cotton profits. Their 7.5 *feddans* of sorghum yielded 45 sacks that year with a labor bill of £S200 (US$250). The sorghum stalks alone are worth £S300 (US$375) to him as fodder for his livestock, he says. He doesn't generally sell any of his own grain. However, he could have sold it for about £S14 per sack at the harvest, £S630 (US$787.50) total, and for £S23 per sack if he waited till the pre-harvest season of the next year. In practice, they eat the sorghum themselves and feed it to the cows for the two months of the rainy season when there is no grazing in the fields. Only if grain remains by the next harvest, does he sell any.

In 1981–82, their 7.5 *feddans* of cotton yielded 22 *kantar*, just under the 3 *kantar/feddan* minimum yield established by the scheme to cover production costs. So again he received no cotton profits. That year their sorghum yielded nothing; birds ate it, leaving only the stalks which he fed to his cows.

Abdelsalaam also rented 40 *feddans* (16.8 ha) of rainland from the government elsewhere in the Blue Nile Province and grew peanuts with hired labor. It yielded 1,000 sacks which he sold to one of the richest traders in Sennar for £S9/sack. That netted £S9,000 gross for the crop, from which he had to deduct plowing costs, labor bills, and a small rent.

Like many villagers, Abdelsalaam still has rights to land along the riverbank (*jeref*) which formed part of the local agricultural system before the scheme. This land is not cultivable since the river no longer rises as it used to because of the Sennar and Roseires dams, the irrigation schemes, and declining rainfall. Abdelsalaam is thinking about making a well there and getting a pump to irrigate the land in order to establish a fruit garden as another source of income.

Hawa and her daughters do no farm work and do not pick cotton. Hawa occasionally makes *tabaqs* which her mother sells for her in the *suq*. These take weeks of intermittent work to make and sell for only five or six pounds. Hawa participates in a rotating credit club (*sanduq*) with neighborhood women, putting in £S2/week. Women often keep some of the daily household food money for themselves, so she probably comes up with the £S2 this way.

Surprisingly, Abdelsalaam says their food expenses are £S4/day (£S120/month), less than Hassan and Fatma's, although this household is bigger. Perhaps the dairy products from their cows and the grain they reap in successful years keep costs down.

Abdelsalaam is an influential man in Wad al Abbas. He is a member of many committees. Almost all the members of committees and boards at Wad al Abbas are wealthy merchants. Villagers assume that they enrich themselves through these positions by diverting money and goods from public purposes to their private businesses. In the case of Abdelsalaam, the extent of his activities in this regard alone suggests that altruistic voluntarism may not be his sole motive. Abdelsalaam's "public service" includes simultaneous positions as president of the local neighborhood council (the lowest level of the *mejlis al shaabi*), president of the building board which deals with local public buildings, president of the land board which allocates land for house-building to villagers, assistant to the president of the board which deals with local schools, treasurer of the *markaz* (government health station) board, member of the farmers' union (a management-sponsored organization of farmers), member of the *lajnat al tenfiziya* (development board) connected with government, and member of the well board that deals with the village's water supply from a government well. He is also involved in the distribution of government-subsidized sugar rations to villagers, and is one of twelve men from Wad al Abbas who sit on the regional board. He is a member of the newly convened Union of Traders, as well. By 1988 he had lost at least one position, the sugar distribution, because of alleged corruption.

The scale of Abdelsalaam's wealth assures that his household will continue to prosper. His sons and son-in-law will assume greater responsibilities in the family businesses and have wealth to estalish their own households and to pass on to their sons.

THESE FOUR households reveal a number of things about economic inequality and agriculture at Wad al Abbas. First of all, none of the households depend solely on farming; even the poorest peasant supplements agriculture with income from petty trade. The wealthiest household has the most diverse economic strategies including trade in cloth, household supplies, and grain, cash crop production on and off the scheme, subsistence production of sorghum for food and fodder on the scheme, and animal husbandry for consumption and sale.

Also significant is the role of institutional connections—the poorest household has no links to scheme or village administration, the poor household that may be making modest gains has connections to scheme administration, while the wealthiest man has a role in virtually every important body in the area.

Neither household size nor stage in the development cycle appear to account for the differences in wealth among these households. While the wealthiest household is the largest, with thirteen members, most of these are dependents; and while it has three working men, so does the household of Osman and Katiira who are not nearly as wealthy. One of the poorest households, that of Hassan, the petty trader, has two working men out of a total household of seven.

In terms of farming, it is important to note that both the poorest peasant and the wealthy merchant employ some agricultural labor. Hired farm labor is not necessarily a luxury. Hassan the petty trader, was able to get by without hired labor only because he cultivated no cotton and left some of his sorghum land fallow. In contrast to the other three households, the wealthy household employs considerable labor not only in agriculture but in trade and animal husbandry as well.

Only the wealthiest man, Abdelsalaam, truly reaped a benefit from his Wad al Abbas land—and even his agricultural success was limited to the sorghum crop in one of the two years. Furthermore, while the two poorest households own no livestock and thus do not benefit from using the sorghum stalks as fodder, Abdelsalaam gains this extra benefit from his land. He spent £S200 (US$250) on hired labor to produce sorghum alone in 1981, while the poorest man, Mustafa, hired no labor, Hassan spent only £S50 (US$62.50) on sorghum labor, and Osman spent nothing on it. Few men in the village have the resources to make the kind of investment in hired farm labor that Abdelsalaam can afford. And even he has little interest in cotton production as evidenced by his heavier expenditures for labor on sorghum, although cotton is a much more labor-intensive crop.

In contrast to Mustafa and Osman, who had lost land, and to Hassan, who gave up sharecropping and left some of his own land fallow, Abdelsalaam has increased his holdings, purchasing land for his son. His wealth and livestock allow him to benefit from scheme agriculture in ways the others cannot.

Most clearly, the size of landholdings is not the root of inequality among these households—until Abdelsalaam's recent purchase, they each owned 15 *feddans* (6.3 ha). It is the difference in their non-agricultural resources that accounts for the variation in the circumstances of their households.

INEQUALITY IN rural communities is often presumed to be based in unequal access to land. The impact of unequal access to off-farm resources on the economic structure of rural communities has received less attention. However, the growing participation of peasants in labor and commodities markets reduces the significance of land in rural political economy. Household size and composition similarly become less important as labor is commoditized. The case of Wad al Abbas is not exceptional. Among the Yoruba in Nigeria, "[d]ifferentiation within the cocoa-growing sector has more to do with farmers' differential links to the regional political economy as a whole than with differential access to rural land and agricultural capital" (Berry 1985:5). On a Mexican *ejido*, economic differentiation means that "[a]vailability of land alone fails to guarantee the peasant equal life chances" (Finkler 1980:283). Income from other sources, moreover, may be converted into agricultural resources.

Including off-farm incomes and non-agricultural assets presents a very different picture of wealth differences than that derived from land ownership alone. Land is not necessarily the major source of wealth even in rural farming communities. Most Wad al Abbas households engage in non-agricultural activities, and access to off-farm resources is crucial in determining the relative prosperity of households and their agricultural success. By looking at the connections between migrants and their rural communities and taking into account the spectrum of activities in which members of the same household are involved, a more complex picture of rural economy emerges. This perspective provides insights into the relationship between peasant household economy and regional and national economic systems. The next chapter focuses on such relationships as they affect agriculture.

NOTES

1. No villager owns a Mercedes but one of the scheme administrators apparently had one that was government-owned. Such perks are not uncommon at the administrative levels of the irrigated schemes.

2. The *umra* is the small pilgrimage that can be performed at any time of year not just during the month of the Haj. Exit visas from Sudan are difficult to get even for religious purposes, but some people use Islamic pilgrimage as a means to enter Saudi Arabia for work.

3. It was not possible to break the households into occupational categories as a measure of wealth. Household members are involved in different occupations, and the incomes associated with a particular occupation vary greatly.

On the survey, wealth was measured in three ways: 1) a qualitative observation; 2) property owned; and 3) reported average daily household consumption expenditures which were divided by the number of household members to get a per capita expenditure which could be compared across households. (This last measure was problematic as an indicator of wealth since larger households appeared to benefit substantially from economies of scale.) While the first two measures were significantly correlated at .702, monthly per capita expenditure was not significantly correlated with either resource ownership or the qualitative assessment of wealth. Therefore it was eliminated from the final measure of household wealth.

4. This points up the difference between father-son relationships where the property is held in common under the control of the senior man and in-law relations where the junior man's separate claim to wealth is maintained.

SIX

The Labor Market and Household Farm Labor

PARTICIPATION IN wage and informal sector employment has altered the dynamics of local agriculture and household farming strategies at Wad al Abbas. Most significantly, it has had profound effects on family farm labor as the market for labor outside of agriculture has come to determine the composition of the household agricultural labor force. The off-farm work and resources of household members influence agricultural strategies in three fundamental ways. First, they provide alternative channels for investment of labor and other household resources. Second, they determine the degree to which the household depends on the farm for subsistence. And third, they affect the level of cash available for inputs to agriculture.

In terms of household labor resources, one effect of the labor market is to raise the opportunity cost of farm labor within the peasant household. In using family labor on the farm, the household is forgoing the rewards that this labor would generate on the wider market. The particular conditions in farming and in non-agricultural employment set the terms of this trade-off. The role of demographic factors and household development cycles in determining the family's agricultural labor force is thus diminished once peasants begin to participate in the labor market. This is often overlooked in studies of farm labor. For example, the ILO (1984) relies on the household development cycle to predict the supply of family farm labor in Sudan's irrigated areas, although they acknowledge that off-farm work and labor migration are widespread.

Off-farm employment opportunities shape the composition of the household agricultural labor force at Wad al Abbas and, by so doing, they affect the range of farming strategies that households can pursue. The allocation of household labor to off-farm work raises the cost of farming since hired farm labor requires cash outlays. For the few wealthy households, whose non-agricultural incomes are substantial, heavy use of hired farm labor is an option and may be the preferred strategy. For the majority of households, peasant-workers whose cash resources are limited, farming strategies that minimize labor requirements are the most viable and attractive. Such strategies enable households to maintain agricultural production despite their simultaneous involvement in off-farm work.

The number of family agricultural laborers supplied by households on the irrigated schemes has been found to average about 1.5 (Kuko 1984). At Wad al Abbas farmers commonly neglect their cotton crops and reports indicate widespread negligence in cotton production throughout the irrigated schemes (Barnett 1977; Khalafalla 1981a). The analysis of the interaction between farming and off-farm work presented in this chapter suggests an explanation for this pattern of behavior and its rationale at the micro-level.

THE LABOR MARKET HIERARCHY

The people of Wad al Abbas participate in national and international labor markets. Their opportunities and conditions of work are thus determined by macroeconomic conditions rather than simply by village institutions or the local supply of labor. It is therefore useful to look beyond the village to the national system into which villagers have been drawn.

Income Differentials Between Farming and Off-Farm Work

The most remunerative work open to villagers is wage-work in Saudi Arabia and other Arab oil-producing states. Trade and transport activities in Sudan also offer high incomes, followed by wage-work in Sudan, petty trade, and, lastly, crafts. Agricultural incomes are the lowest of all. Such income differentials between agriculture and off-farm work are common in Africa and throughout the developing world. For example, in Nigeria, "Even in colonial times, although the spread of cocoa growing and commercial food crop production created widespread opportunities for modest increases in income, the best opportunities for accumulation lay not in agriculture but in trade, the professions, and the civil service" (Berry 1985:12).

A 1974 ILO survey of incomes in different occupations in the Three Towns (Khartoum, Khartoum North, and Omdurman) classified occupations into eight groups, plus an additional miscellaneous category. On the basis of the data they present, it is possible to determine the median monthly income of each occupational category for their sample (ILO 1976). The results are shown in table 6.1. Those employed in agriculture are among the lowest paid, along with unskilled workers and service workers. The median income of clerical workers, for example, is twice that of those in agriculture. Administrative, professional, clerical, and trade-sales workers are the highest paid. Production workers and those not included in the categories fall in between.

In 1974 average annual incomes for a sample of 443 employed migrants to the Greater Khartoum area were £S217 (US$622.79)[1] in services, £S251 (US$720.37) for unskilled workers, and £S273 (US$783.51) in manufacturing/production (ILO 1976:356). Significantly, these incomes "are far above what they could earn in agriculture" (ILO 1976:356). The ILO also concludes: "We have several good reasons to believe that the commerce sector occupies a privileged and protected position in the economy . . . [O]pportunities to make easy profits in the commerce sector have drawn in resources (especially entrepreneurial talent) from the pro-

TABLE 6.1. Incomes in Various Occupations in the Three Towns, 1974

Occupational Category	Median Monthly Income		N
	£S	US$	
Administrative	83.33	239.16	2,400
Professional	56.20	161.29	17,050
Clerical	45.79	131.42	28,350
Trade-sales workers	37.46	107.51	24,200
Production workers	27.04	77.61	75,250
Agricultural	22.88	65.67	5,500
Service workers	22.88	65.67	30,900
Unskilled laborers	18.71	53.70	8,050
Other	27.04	77.61	10,700

SOURCE: Adapted from ILO 1976:341, table 86.

NOTE: £S1 = US$2.87 in 1974.

ductive sectors, and slowed the development of the latter" (1976:452–453).

A smaller study conducted in 1982 in Greater Khartoum and Port Sudan also found great gaps between agricultural[2] and non-agricultural incomes (ILO 1984). Forty-four percent of those employed in agriculture earned £S33.33 (US$37) or less per month. In all other types of employment, the percentage in this income group did not exceed 8 percent, which was for service workers. Sixty-eight percent of professionals earned between £S200.08 (US$222.09) and £S500 (US$555) per month, while 43 percent of managers earned that much. The majority of skilled workers, 52 percent, averaged £S83.42 (US$92.59) to £S200 (US$222) per month. High incomes in commercial activities are reflected in the fact that of those categorized as employed in "own trade," 82 percent had earnings over £S83.33 (US$92.58) per month, and 60 percent earned over £S125 (US$138.75) per month (ILO 1984:204, table A18). This study also noted that for Sudanese international migrants "the rates of pay in the host countries are 6–10 times the pay for equivalent work in Sudan" (ILO 1984:150).

A 1981 survey found mean agricultural incomes of household heads to range between £S15.92 and £S77.17 per month in six villages on the Gezira Scheme (ILO 1984:107). These villages include some composed mainly of agricultural laborers and sharecroppers, as well as some composed mainly of tenants. Mean agricultural incomes among household heads in the tenant villages range between £S56.83 (US$71.04)/month and £S70.33 (US$87.91)/month. Most Wad al Abbas farmers had no farm income that year. Conditions are better on the Gezira because the gravity irrigation is more plentiful and reliable than pumped water and holdings are larger, enabling farmers to grow cash crops such as wheat, peanuts, and vegetables in addition to cotton and sorghum. Even the highest of these Gezira incomes is low, however, compared to the pay for most non-agricultural work.

Wad al Abbas villagers are engaged in commercial activities that vary widely in scale of operation, capital, and income.[3] Since trade profits often depend upon price differentials between regions or between urban and rural areas, transport is an essential component of trading and a natural area of expansion for successful traders. Some of the big merchants of Wad al Abbas have made enough money to purchase their own trucks. One says he paid over £S12,000 in cash for a truck in 1977. Two middle-aged married brothers, Ali and Al Nil, who trade together as partners, ordered a truck from England for £S20,000 in 1980 and had it shipped to Sudan. Al Nil says their business is worth £S50,000. In the fall of 1980 he was planning to fly from Khartoum to Juba and bring goods north by river barge, since no trucks could get through because of heavy rains there. They regard farming as work for the poor. As Al Nil explains:

> We have an older brother, but he has no money. I tired myself out. I went to the west and El Obeid, Nyala, and tired myself out [trading] and brought back money. He didn't leave the village. He takes care of the tenancy. But I give him money . . . . A few years ago we were all poor. We used to go to the fields and eat *kisra bil moya* [a meal of bread moistened with water]. We just made enough to eat. You work in farming all year and don't get a *girish* [cent]. . . . Ali was a janitor in the school, then I came back from the west and we started business. Now trade is going well. We both gave our tenancies away and don't even ask money for it, just let someone farm it and eat from it. I gave mine to my *amm* [father's brother].

The work of another villager illustrates the profits in trade and transport. In 1981 he was buying 100 sacks of dates for £S46 each in Khartoum and selling them for £S50 each in Wad al Abbas, making £S400 (US$500) per trip. Traders without trucks do the same thing, but must arrange and pay for transport. One buys *towb*s for £S110 in Khartoum and sells them for £S120 to traders in Sennar who retail them for £S130. Another takes peanut oil from Sennar, where he pays £S20.25 per container, to Kosti, where he sells it for £S21 per container, making £S200 a truckload.

At the other end of the spectrum from the big merchants are the many petty traders, with no property and little capital, who barely manage to eke out a living on a few pounds profit per day. They buy and sell in small quantities and have a much slower rate of turnover than bigger traders (who generally sell wholesale to smaller traders). Some petty traders buy vegetables in Sennar and sell them for a little more in the village. One buys saddles in Wad al Abbas for £S5 or so and sells them for about £S7 in Sennar. Another buys 100 *tabaq*s (fiber tray covers) in the village for £S5 each and gradually sells them in Kosti for £S7 each. Others are a little better off. For example, two men, working as partners, make £S150 each per month buying kerosene in Port Sudan and selling it elsewhere.

Most Wad al Abbas traders are small retail sellers. Traders are secretive about their cash flow, so it was not always possible to get precise income figures. By my estimate, the great majority of traders fell into the £S100 to £S200 (US$110–$222) monthly income range in 1982. One of the most prosperous merchants said he could make this much in a single day. The trading activities of most villagers can be considered part of the informal sector in terms of being small-scale operations of the self-employed. But even large-scale traders are part of the informal sector in the sense that much of their activity is not documented or taxed, and they engage in black market activities such as selling government price-controlled commodities above the official price. According to one petty trader, "Those are the only things you can make money on. You get a quota from the government and you're OK." A few petty traders and craftsmen only work seasonally so that they can continue to farm, closing up shop or stopping work during the rainy season when crops need the most attention.

Passenger transport is less lucrative than transporting goods to sell, but some villagers say they don't have the knack for trade. Passenger transport offers a fairly steady income, although the price and supply of fuel are often problematic as are vehicle repairs and spare parts. Several villagers own trucks that have been converted to buses which they run between Wad al Abbas, Sennar, Medani, and other routes, and a few own taxis which they operate in towns. One villager with a bus was grossing £S100 (US$111) round trip in fares, transporting passengers between Sennar and Medani in 1982.

The scale of wages and salaries for different jobs and levels of training is fairly standardized throughout Sudan. The earnings of wage-workers from Wad al Abbas ranged from about £S35 (US$38.85) per month for the lowest paid full-time work to £S160 (US$177.60) for a university graduate with a good job, between 1980 and 1982. The director of a school in the Northern Province was earning £S150/month while another villager, employed as a health worker in the Blue Nile Province, earned £S160/month. A young villager, working as a casual laborer in government construction in Gedaref, earned £S6 a day, when employed.

Men sometimes supplement income from low-paid wage-work by engaging in trade on the side. The fact that government business hours end at 1:00 p.m. for the day (due to the heat) facilitates this. If a man is successful enough to build up capital, he may leave his job to trade full-time. As one man expressed a common sentiment: "Trade is good. There is nothing in government work."

For the Wad al Abbas men employed in Saudi Arabia and other oil states, salaries ranged from the equivalent of £S400 (US$444)/month to more than £S1,000 (US$1,110)/month, often with fringe benefits, such as housing and transportation, that allow migrants to accumulate savings. A villager who drove a truck in Saudi Arabia was earning £S800/month in

1981. As one villager remarked, 'There's no money in work here [in Sudan], that's what's brought us all this migration." When a migrant returns to the village from abroad, he is greeted like a king. His family celebrates for days and sacrifices a sheep to feast neighbors and kin. International migrants generally bring suitcases full of gifts for their extended families and token offerings, as little as a bouillon cube, for more distant guests who drop by. An old woman observed: "Those who go to Saudi Arabia, their money, they spill it on the street. They have everything. The one who didn't go has nothing and what he does have is broken or no good."

Crafts are the least lucrative non-farm work in which villagers engage. A craft at Wad al Abbas is usually practiced by an individual working alone. One exception to this is a man who has organized several hired workers into a shoemaking factory of sorts. The men work in tandem, one cutting leather, the other sewing it. They each get 50p per pair of shoes and are able to make 6 pairs (£S3) in a day. One advantage of craftwork is that it can be carried out in the village, allowing craftsmen to remain with their families. For this reason it is also compatible with farming in a way that most other work is not, with the exception of petty trade within the village. From the statements of various craftsmen regarding their productivity, costs of production, and sales proceeds, their monthly income ranged from £S60 (US$66.60) to £S100 (US$111) for *full-time* work in 1982. However, most of them are not able to sustain a consistent level of activity. Many craftsmen have other responsibilities such as farming or, in a few cases, attending school. Also, many depend on demand within the village itself, which is limited, though some sell their products in Sennar or Wad Medani. Because of limited demand and inconsistent production, monthly incomes ranging between £S30 (US$33.30) and £S50 (US$55.50) in 1982 were common.

Villagers prize wage-work for the relative security it offers compared to self-employment, but they dislike the inflexible work schedule. Much wage-work is in the public sector and villagers call it *"shughul al hukuma"* (government work). They contrast it with *"shughul al hurr"* (free work) by which they mean trade and other forms of self-employment. Villagers recognize that while incomes in trade and crafts fluctuate, wages and salaries are steady but limited. A villager teaching in the local school says: "Here, especially if you work for the government it doesn't matter how you work, your salary will come just the same. And, whether you work well or not, you won't get a lot of money." Not long after that he left to work abroad as a teacher. Another man left his post on the Sennar police force to try to trade and then left that shortly to look for work in Saudi Arabia.

Trade, unlike wage-work, offers the possibility of amassing great wealth. While this possibility is remote for the majority, some villagers have succeeded. Most of the wealthy merchants in the village today did not inherit their wealth but accumulated it over the last three decades. Some of them even continue to trade in the same basic commodities—spices, oil, onions, etc.—with which they began, but in much greater volume. In 1982, one village trader was planning to open a store in Omdurman with goods worth £S5,000 to start. As his daughter tells it,

> First he was just a plain farmer, but when he saw the tenancy doesn't give enough money he left it to be sharecropped and began to trade. Little by little [over the last twelve years] he built up capital. The scheme took away our land [for negligence] but gave it back under condition that it be registered under a different name, so now it is registered to [my brother].

Men leave most occupations for trade if they can. The more lucrative their work, the higher they can start out in trade. Even educated men with prestigious jobs in the capital leave these positions to go into trade. There were several instances of this in the village between 1980 and 1982, including at least one university graduate. There was even one villager who broke down part of his house and sold the materials to raise cash for trading. Migrants who return from abroad to work in Sudan generally enter trade or transport if they have been able to accumulate capital. A migrant who had been abroad for three years was rumored to have saved £S19,000 (US$23,750) when he returned to the village in 1981.

One verse of a popular girls' chant about desirable husbands says: *Wa la mughterib wa la tajir fil rarib* (either an international migrant or a trader in the west). Western Sudan is an area of commercial activity by traders from northern and central Sudan (*jellaba*). Another verse goes, "Either a migrant for two years or an officer with two stars." When on one occasion a woman joked, "What about a farmer with two tenancies?" all the girls present laughed. Villagers equate farming with poverty. As one man said, "I am 54 years old. I just stayed on the land. We eat *kisra*, that's all."

Preceding chapters have established the extremely low incomes in agriculture at Wad al Abbas. Cotton is rarely profitable and sorghum production seldom meets household consumption needs, let alone yields commercial surpluses. Virtually all forms of non-agricultural work yield higher incomes than farming in Sudan. Work abroad, white-collar work in Sudan, and trade are the better paying types of employment. Villagers are well aware of this. Thus, rationally, they prefer such work to farming if they can get it. The number of men involved in trade and wage-labor is one indication of this.

Given the higher returns from off-farm work and the low profits and high risks of farming at Wad al Abbas, one might ask: why do any villagers farm at all? This is the reverse of the usual formulation, "why do peasants migrate for wage or informal sector employment?" As we have seen, by the early 1980s, many individuals had left farming. But if conditions are as bleak as this research indicates, why do households continue to farm at all?

WHY DO THEY FARM AT ALL?

The answer lies in the interaction of the labor market and household economy. The labor market assigns different values to different kinds of labor—not all labor-power is equally rewarded or equally saleable. The household, in contrast, supports all its members, even the very young, the aged, and the ill, who cannot labor. It is through the interaction of these two systems that farming takes the form it does at Wad al Abbas. Farmers are often those who have no market outlet for their labor and who are economically dependent on other household members.

Up to now, I have described the various occupations of Wad al Abbas villagers and the different incomes associated with them as if they were alternatives to farming. But analyses that simply focus on options fail to address the constraints under which people operate. In considering the role that involvement in wage-labor and commerce play in peasant communities, it must be recognized that access to wage employment or commercial enterprise is not evenly distributed. Skills, education, capital, and, sometimes, connections are prerequisites, to various degrees, for different occupations.

The attempts of Wad al Abbas men to find jobs and to migrate abroad are not always successful. Migration even within Sudan requires transportation, food, and housing away from home. Jobs are not easy to come by and waged or salaried jobs generally require some education or skill. Periods of unemployment lasting six months or more are not uncommon. Some men's search for work is only intermittently interrupted by brief periods of employment. Moreover, in the early eighties the Sudanese government was trying to control rural-urban migration and conducted periodic sweeps of the capital and major towns, arresting people without local identification cards or evidence of employment and holding them in camps outside the city or forcibly transporting them to rural areas. Villagers were well aware of these raids and justly fearful. In fact, several village men were arrested this way in Khartoum, but their release was secured by relatives there.

Migration abroad is particularly costly and difficult, often entailing both the use of personal contacts and large cash expenditures. As one woman remarked, "If it weren't so hard to get papers and visas, all of the men would be over in Saudi [Arabia]." Young's (1987) work on Rashayda labor migration from Sudan similarly shows that considerable resources are required to launch a migrant. In 1981 Wad al Abbas villagers were paying £S200 or even £S250 (US$312.50) to obtain exit visas from Sudan, and some paid as much as £S300 (US$375) to apply for a Saudi entry visa. Airfare was £S125 roundtrip. Villagers also pay dearly to obtain foreign work contracts through the black market in Khartoum. Some men use the *umra* and the *haj* to enter Saudi Arabia, and then look for work there illegally, in some cases returning to Sudan with the contract and re-migrating through official channels. Sometimes men simply fail and return home empty-handed.

Trade, especially petty trade, is much easier to enter but not all who enter trade are able to make a living at it. To be a successful trader requires capital or at least access to credit, which, in turn, is predicated on the ability to repay. It may also require a certain business acumen and personal connections that some villagers lack. While those who can, leave other work for trade, unsuccessful traders look for steady wage-work. For example, one young man who had been trading as a *waggaafi*, getting and selling goods in different places, left that to look for work in the capital because, as his mother explained, "He's tired of that, he wants a job sitting around in one place with a pen."

In Sudan's labor market, education is a fairly good predictor of occupation and income. "All the evidence we have reveals a strong positive correlation between education and earnings from work" (ILO 1976:358). The ILO estimated that a university graduate entering the labor market in 1974 would be "paid two to three times as much as a primary-school leaver" (1976:129). Table 6.2 shows the association of occupation and education in Sudan in the mid-'70s. Among those employed in agriculture, 66.4 percent had no formal education while 89.6 percent of those in administrative occupations had secondary school education or higher, as did 58.7 percent of those in professional occupations. Seventy percent of unskilled laborers had no formal education as compared to 48.5 percent of those in trade-sales, and 2.3 percent of clerical workers. Of 8,050 unskilled laborers, none had completed secondary school. Only 3.6 percent of those employed in agriculture had secondary education or higher. The educated obtain jobs with higher incomes; the uneducated are kept in low-income work.

This remained the case through the 1980s. A smaller 1982 study of urban Sudanese labor markets found a strong association between education and occupation (ILO 1984). Eighty-one percent of professional and white collar workers had secondary school education or more, while only 17 percent of skilled workers and 9 percent of unskilled workers were educated to this level (ILO 1984:207, table A21). Six percent of professional and white collar workers had only primary education or less compared to 62 percent of skilled and 74 percent of unskilled workers.

Uneducated Sudanese are clearly at a disadvantage in terms of access to the higher-paying administrative, professional, and clerical jobs. Farmers, agricultural laborers, and unskilled workers are among the lowest income groups and are the least educated.

In terms of household farming, the higher incomes outside of agriculture can be seen as an opportunity cost of using family labor on the farm. But not all labor is remunerated equally and therefore it is associated with different opportunity costs in peasant-worker household economy. In the

TABLE 6.2. Levels of Education in Various Occupations in the Three Towns, 1974

Occupational Category	Percentage with No Formal Education	Percentage with Secondary School Completed or Higher	N
Administrative	—	89.6	2,400
Professional	1.8	58.7	17,050
Clerical	2.3	51.8	28,350
Trade-sales workers	48.5	5.3	24,200
Agricultural	66.4	3.6	5,500
Production workers	33.7	2.2	75,250
Service workers	63.3	1.2	30,900
Unskilled laborers	70.2	0.0	8,050
Other	34.4	5.8	10,650

SOURCE: Adapted from ILO 1976:337, table 83.

competitive labor market, the uneducated and unskilled have difficulty marketing their labor. Labor which is not saleable has little opportunity cost in family farming. If not employed on the farm, it would otherwise be idle.

National and even international labor markets thus have a powerful impact on peasant farming. When peasant households cannot support themselves through agriculture alone, they seek to market their labor elsewhere. In the competitive labor market, the more skilled, educated, youthful labor finds wage or salaried employment, or enters commerce. The rest of the population is effectively marginalized. The capitalist economy cannot absorb their labor directly. This "surplus" labor, then, by its characteristics is not so much a reserve industrial labor pool as it is a reserve labor pool for the least productive, least developed sector of the economy — peasant agriculture.

Not only are skills and education important determinants of employment opportunities and income in Sudan, gender also plays a significant role. Wad al Abbas women have not had equal access to the education and training that would qualify them for employment. Furthermore, there is discrimination in wages and sex-typing of jobs in the Sudanese labor market and the international markets that employ Sudanese abroad (ILO 1984). The market for female labor is thus much more limited than that for male.

At Wad al Abbas, women's labor is not marketed for the most part, with the exception of local cotton harvesting, and more recently, the few female high school graduates who work as teachers. Cotton picking is seasonal, arduous, low-paid, low-prestige work. Women in families who can get by without it do not pick cotton. Thus, women's labor has no opportunity cost for the Wad al Abbas household and its allocation has not been directly affected by the labor market. Rather women's activities and labor time are affected indirectly by household strategies for allocating male labor, since women's domestic work and uninterrupted residence in the village facilitate male employment outside the household and the community. This division of labor by sex, in turn, presents an obstacle to women's participation in work outside the home or the village.

The withholding of women's labor from the market is reinforced by cultural norms of sex segregation and female seclusion. The spread of orthodox Islamic ideologies in Wad al Abbas has played a role in legitimizing the increasing dichotomy of gender roles that has emerged from the masculinization of agriculture by the irrigation scheme and from male labor migration. But Islam is not the causal factor. As Kapteinjs (1985:57) has argued in the case of Dar Fur, "The position of women . . . has not deteriorated because people have become better Muslims but because—in the twentieth century—the area has come to be dominated by a new economic system which has taken a more literal observance of (certain rules of) Islam as its ideology."

Gender roles in Wad al Abbas are complex and changing. Evidence suggests that a withdrawal of female labor from low-status public work such as agriculture has taken place since the establishment of the irrigation scheme (Bernal 1988a). Only since the late '80s are there indications that this may eventually be followed by a movement toward marketing women's labor in higher paying, higher status areas, notably teaching. Nonetheless, most households currently lack women with the education to qualify for these jobs. Thus, despite some evidence that women's labor, too, is becoming a commodity, as of now the labor marketed by Wad al Abbas households remains almost exclusively male.

Most Wad al Abbas women and girls do not engage in paid work and take no part in farming operations besides cotton picking where they work for tenants as hired laborers. Women are thus not part of the household agricultural labor force, although as cotton pickers they are an important part of the scheme's production system. It is males who cultivate household land as unpaid family labor and/or market their labor off the farm. The following discussion therefore focuses on male labor.

OFF-FARM WORK AND HOUSEHOLD FARM LABOR

Age and education are important determinants of a Wad al Abbas man's occupational options. The two are related because educational opportunities were limited in the past. The British attempted to establish a school at Wad al Abbas as early as 1912. But, by their own account, villagers opposed the school and it was soon closed. The contrast between that early reaction and the present interest villagers have in education is telling. Public schooling had little value when children were going to remain and make their livelihood in the agrarian community. Now that parents see their children's fate (and their own in old age) as tied to the national economy, where education is a criterion for access to the higher paying jobs, schooling is important to them. Villagers are currently pushing the government to establish a high school in the village.

Until 1957 Wad al Abbas had no government school, only the Islamic *khalwa* established at the founding of the village. A few local boys attended elementary school in nearby villages. In 1957 a boys elementary school was built at Wad al Abbas. Even then boys who wished to complete their elementary education had to study in other villages as the Wad al Abbas school initially went only to third grade. Boys are generally eight to ten years old when they enter elementary school at Wad al Abbas, so men born before 1949 had little opportunity for even elementary education.

It is no coincidence that a school was established at Wad al Abbas shortly after the irrigated scheme. Villagers say they were impressed by the educated scheme employees, and that the cotton profits of the early years stimulated them to undertake self-help projects like the school. A girls elementary school was later built in 1963 and intermediate schools were established in 1966 and 1973 for boys and girls, respectively. Villagers must still go outside the village, usually to Sennar, to attend high school.

Table 6.3 shows the educational level and occupations of men in the random sample of fifty-three households surveyed in 1982. In order to obtain the largest possible sample of men in different occupations, all 53 households are included, although nine of them were landless at the time of the survey. Six years of education is a logical breakoff point because that is the duration of elementary school. There are a total of 108 men in these households of whom 86 (79.6%) are employed or self-employed. Lack of data on the education of some of these men reduced the sample size to 64 men. The percentage of elementary school graduates is highest among wage-workers (54%), then traders (52%), followed by craftsmen (40%) and farmers (0). Most significant is the fact that none of the 15

TABLE 6.3. Education of Wad al Abbas Men in Various Occupations

Occupation	Total Men	Men with Six Years or More Education	
		Number	Percent
Wage and salaried workers	13	7	54
Traders	31	16	52
Craftsmen	5	2	40
Farmers	15	0	0
	N = 64		

NOTE: Data on men in 53 households.

farmers whose educational level was known had completed six years of education. In fact, none had completed more than three years of schooling. Furthermore, 93 percent (13 out of 14) of the farmers say they can neither read nor write, while 90 percent of the traders (28 out of 31) say they can.

A study of Gezira Scheme households similarly points to the importance of education in determining off-farm work opportunities—78.9 percent of migrants are literate while among the majority who remain in their village, only 38.6 percent are literate (Yousif 1985:91).

Table 6.4 shows the age distribution of Wad al Abbas men in various occupations. It reveals that farming is the work of old men; 73.7 percent of farmers are 50 or over. Wage and salaried workers, at the other extreme, are young. Most (87.5%) of them are 30 or younger. The association of education and wage-employment means the virtual exclusion of older men from these jobs. The high average age of farmers, however, not only reflects the lack of education and training received by this generation but men's work trajectories. As men age they move from being income producers to dependent household members; they retire from other work into farming and are supported by their sons. Farming allows elder men to remain permanently in the village and they take up the anchoring role, overseeing family property and social relations, while junior men are away from the village working.

Like wage-workers, many traders are also young men: 54.5 percent of traders in the sample are 30 or younger. But, unlike wage-workers, only

TABLE 6.4. Ages of Wad al Abbas Men in Various Occupations

Occupation	Age							
	30 or Younger		35–45		50 or Older			
	#	%	#	%	#	%	N	%
Wage and salaried workers	14	87.5	1	6.25	1	6.25	16	100
Craftsmen	3	50.0	2	33.3	1	16.7	6	100
Traders	18	54.5	5	15.2	10	30.3	33	100
Farmers	2	10.5	3	15.8	14	73.7	19	100
							N = 74	

NOTE: Ages are estimates rounded to the nearest 5 years.

6.25 percent of whom are over 50, over 30 percent of the traders are in this age group. Fifty percent of the craftsmen are in the youngest age group, a third are between 35 and 45, and only one-sixth are 50 or over. However, since the number of craftsmen sampled is so small, it is not clear to what degree this distribution is representative.

To summarize, the average wage-worker from Wad al Abbas is young, literate, with at least an elementary education. Four (31%) of the 13 wage earners whose education was known, had a high school education or more. This corresponds with research at the national level showing labor migrants to be predominantly young males with more education than the general population (Kuko 1984; Mohamed 1986). The average farmer is old, uneducated, and illiterate. Traders fall somewhere between these extremes. Their ages are broadly distributed; and while a greater proportion of traders have elementary education than do farmers, traders have less education than wage-workers. Only 4 (13%) of the 31 traders whose education was known had a high school education.

Since the majority of farming households at Wad al Abbas have more than one employed or self-employed male, the distribution of men across various occupations is reflected within the household as a division of labor among its members. In this way, the labor market has a direct effect on peasant household economy and the composition of the agricultural labor force.

Peasant producers are known to work for very low returns since they must keep working to satisfy their own needs, regardless of diminishing returns. If the peasant household were an entirely self-provisioning unit of production and consumption, the intensity of its labor would be governed solely by these internal needs. Once peasants satisfy some of their needs for subsistence or agricultural inputs through the market, they may be forced by external pressures to intensify their labor to satisfy their basic needs. Their capacity for self-exploitation thus becomes a means through which value is transferred from the peasant household to the consumers of peasant products, including labor. Unpaid family labor and the production of use-values for own consumption contribute to the reproduction of the household and subsidize the products it sells on the market, allowing them to sell below their cost in terms of the labor they embody.

At Wad al Abbas people are not entirely unaware of this. Discussing sorghum production, a farmer exclaimed, "If we actually counted [the cost of] our own labor, it would be cheaper for us to buy sorghum than to grow it ourselves!" But, if peasants themselves see that it is cheaper to purchase sorghum than to grow it, why *do* they grow it? Because they are under pressure to intensify their labor and some of their labor has no alternative productive outlet. Wad al Abbas households do not impute wages to the labor that produces sorghum since this labor has no value on the labor market. To meet their needs, peasant-worker households allocate labor to farming, despite low returns to labor, because they possess some labor that cannot be sold for wages. In the Wad al Abbas household, this labor is the labor of older, uneducated men, and children.

Wad al Abbas households have become increasingly dependent upon purchased goods, and they face rising costs of food, clothing, and other necessities. Says one villager, "Before, people here had few things, and they were cheap. Now we have many things, but they require a lot of money." Not only has Sudan experienced high rates of inflation over the 1970s and '80s, but most goods, including agricultural products, cost more at Wad al Abbas than they do in towns. This is because goods are brought to the village from outside, entailing additional costs for middlemen and transportation, and because government-subsidized commodities are sold at high black market prices in the village, due to the comparative lack of official oversight in the countryside.

A local school administrator from outside the village explains: "Everything is more expensive [in Wad al Abbas]. The traders make agreements with one another. They buy tomatoes in Sennar and agree on the price here. And here they are all related. No one can report his *wad amm* or *wad khaalat*; it is shameful. The villages in Sudan are all like that." Many villagers complain about black marketeering and corruption among local traders, but they say no one would come forward to support someone who informed the authorities on a fellow villager.

In response to the rising cost of living, households intensify their labor output by having some members employed off the farm while nonetheless struggling to maintain household farming.

THE LOGIC OF HOUSEHOLD FARMING AT WAD AL ABBAS

In terms of profitability and returns to labor, agriculture at Wad al Abbas is unrewarding. But from the perspective of peasant-worker household economy, the returns look different. If the labor of one man is counted as 26 person-days/month and a boy as 13, their labor totals 39 person-days/month.[4] Only in one month, November (the harvest), does the labor required to cultivate 5 *feddans* (2.1 ha) of sorghum exceed this (see table 6.5.) In fact, for every month but November, the labor required on 5 *feddans* of sorghum is less than 26 person-days/month or one man working full-time in agriculture. One man or a man and a boy could cultivate 5 *feddans* of sorghum with little hired labor. But, of course, on the scheme, at least minimal efforts must be made in cotton production as well, to avoid losing the land.

However, if the household relinquishes hope of cotton profits and merely performs operations well enough to pass inspection, it can conserve labor on cotton. If by neglecting cotton, a household reduces labor inputs to cotton by half, a man and a boy (39 person-days of labor/month) would be sufficient to cultivate 5 *feddans* (2.1 ha) of sorghum *and* 5 *feddans* of cotton except for the two harvests (see Table 6.6).

All cotton pickers are paid, but since they are paid by volume, costs are related to yields. The negligent household's yields are likely to be lower, and they may also keep picking expenses down through insufficient picking. Assuming picking labor is also halved, 104.4 person-days of labor must be hired from January to April (table 6.6). (Average picking costs at Wad al Abbas are difficult to calculate because pickers are paid by volume.[5] Pickers work a much longer day and earn less than other

TABLE 6.5. Average Labor Required to Cultivate 5 *Feddans* Cotton and 5 *Feddans* Sorghum

Month	Person-Days/5f Cotton	Person-Days/5f Sorghum	Combined Total
January	53.5	.5	54.0
February	76.1	—	76.1
March	57.3	—	57.3
April	21.6	—	21.6
May	31.5	.2	31.7
June	4.6	6.2	10.8
July	4.1	23.8	27.9
August	27.5	13.4	40.9
September	27.0	4.9	31.9
October	11.5	10.4	21.9
November	6.1	51.7	57.8
December	7.5	16.2	23.7
Yearly Total	328.3	127.3	455.6

SOURCE: ILO 1976.

agricultural laborers.) If pickers earn an average of £S1/person-day of labor (at 1981 rates), picking costs would come to £S104.40.

If one man and a boy are available for farming a tenancy, the only other time hired labor is required is for the sorghum harvest in November. By these calculations, the household would fall short by 16.2 person-days of labor that month. Harvesters earned about £S2/day in 1981, so 16.2 person-days come to £S32.40. The £S104.40 for cotton picking plus £S32.40 for harvesting sorghum total £S136.80. However, over the course of the agricultural year, the scheme provides £S136 cash credit[6] for cultivating and harvesting 5 *feddans* of cotton. The household must also hire a tractor for plowing the sorghum land (£S10 at 1981 rates) and spend a small amount for seeds. Under these conditions, the household's actual cash outlays for production costs on one tenancy come to only about £S11 (US$13.75) for the year (see table 6.7).

In a good year, 5 *feddans* of sorghum can yield 25 or 30 sacks. In October 1981 before the harvest at Wad al Abbas, sorghum was selling for £S23/sack. At this price, 30 sacks are worth £S690. Generally households at Wad al Abbas do not sell grain unless surplus remains when the next harvest is near. Most Wad al Abbas households are *purchasers* of grain rather than sellers, and the average Wad al Abbas farmer does not pro-

TABLE 6.6. Average Labor Required to Cultivate 5 *Feddans* Cotton and 5 *Feddans* Sorghum, if Cotton Labor Is Halved

Month	Person-Days/5f Cotton, Halved	Person-Days/5f Sorghum	Combined Total
January	26.8	.5	27.3
February	38.1	—	38.1
March	28.7	—	28.7
April	10.8	—	10.8
May	15.8	.2	16.0
June	2.3	6.2	8.5
July	2.5	23.8	26.3
August	13.8	13.4	27.2
September	13.5	4.9	18.4
October	5.8	10.4	16.2
November	3.5	51.7	55.2
December	3.8	16.2	20.0
Yearly Total	165.4	127.3	292.7

SOURCE: Based on ILO 1976.

duce sorghum in order to sell it. Many households do not even achieve self-sufficiency in grain with any regularity. But sorghum is expensive, and households can conserve cash by producing grain even with low returns to labor and poor yields, provided they rely on unmarketable household labor to produce it.

When the household runs short of grain before the harvest, prices are at their peak, and villagers try to avoid purchasing grain, or purchase as little as possible. One way they do this is through harvesting grain little by little for daily use as it ripens, before the whole crop is ready for harvest. They also buy grain in small amounts, by the *kayla* (27.7 lbs.) or less. A *kayla* was selling for £S4 in October 1981, bringing the equivalent sack price to £S30. (Prices drop substantially after the harvest.) An expense of £S11 or so for the cultivation of five *feddans* of sorghum is a very small amount compared to the cash expenditures and incomes of Wad al Abbas households. A total yield of one or two sacks would cover the cash investment, and the land will have been secured for another year.

Holding onto land, which at Wad al Abbas means keeping at least the cotton portion under cultivation, is in itself a hedge against the uncertainties in the national economic system and the risks and instability of non-farm income and employment. Villagers often retire into farming when they are too old to pursue other work, and they pass the land down to

TABLE 6.7. Costs of Production on a Tenancy with One Man and a Boy, if Cotton Labor Is Halved

Input	Cost @ 1980–81 rates	
	(£S)	US$
Family labor	No cost as it would be supported anyhow	
Hired labor:		
Cotton picking	£S104.40	$130.38
Sorghum harvest	£S32.40	$40.50
Tractor rental	£S10.00	$12.50
Sorghum seeds	unknown	
Sub-Total	£S146.80	$183.38
Minus Scheme Credit	−£S136.00	−$170.00
Total Expenses:	£S10.80	$13.38

NOTE: £S1 = US$1.25 in 1981.

their heirs. Moreover, as one stated: "That land is our land, land of our grandfathers. Can a person leave it? Or see it given to others and keep quiet? We can't leave that land. It is ours." Clearly, more than ancestral and emotional ties keep peasant-workers attached to their land. Given the conditions they face as sellers of labor in the market and as purchasers of food and other necessities, farming remains an important component of household economy, helping them to survive on small cash incomes.

The preceding model accounts for the most common pattern of farming at Wad al Abbas, that is: minimizing labor inputs to cotton, attempting to produce a successful sorghum crop, and relying primarily on the unpaid labor of an older man and/or children. While household landholdings, labor resources, and off-farm incomes vary, the majority of Wad al Abbas households operate under similar constraints. The necessity of combining off-farm employment and household farming reduces the unpaid household labor available for farming. Only the minority of wealthy households with high off-farm incomes can afford substantial expenditures on hired farm labor if they choose. Most households don't have this option. Over 60 percent of households in the sample are poor (with wealth scores of 2 and 3); they cannot afford to hire much labor or to forgo the chance to produce some of their own food. Thus, if they have more land or less labor than the one man and a boy in this example, they may have to further reduce labor to cotton or even to sorghum, use sharecroppers for part of their holdings, or even leave some of the sorghum land fallow.

This analysis illustrates how important it is to view peasant farming not only in the context of household economy, but in the context of regional economy.

THE IDEA that subsistence agriculture plays a role in reproducing wage laborers and can thus be said to subsidize capitalist production has become common currency (Meillassoux 1981; Goodman and Redclifft 1982). The other side of the coin is that once the peasant household relies heavily on purchased foodstuffs and/or purchased inputs to agriculture, subsistence production itself may be subsidized through commercial and wage-labor activities. Murray, writing on Southern Africa, makes the significant observation that "agriculture and [labor] migration are not in fact alternatives, since agricultural output partly depends on cash investment from migrant earnings" (1979:342). Labor migration from farming communities and remittances to them are widespread phenomena. This example shows how they can play a major part in transforming agricultural production in the labor-sending areas.

At Wad al Abbas, the dependence of the farmer and his household on other sources of income permit them to continue farming even though the returns from agriculture are uncertain and often fall below subsistence levels. For Sudanese peasants and pastoralists generally: "The availability of wage-labor has given sub-reproductive units a new ability to maintain production due to the income earned by family members elsewhere" (O'Brien 1980:261). For example, in a Takari (Nigerian Sudanese) village west of Sennar: "The domination by capitalist production over earlier forms of household production is such that sons engaged in the former invariably have to subsidize the small-scale farming activities and casual employment of their fathers, thus effecting a complete transformation of the relations of personal dependence within the household" (Duffield 1983:57).

But the fact that farmers are supported as dependent household members has consequences far beyond the changes in "relations of personal dependence" cited by Duffield. For one thing, as I have pointed out, returns from agriculture are free to fall below subsistence level as the agricultural labor force is reproduced in part through other relations. Furthermore, the maintenance of household agricultural production becomes dependent upon participation in capitalist relations of production and exchange. Not only does off-farm work help to sustain household farming, but the low incomes from such work and the instability of national markets compel households to maintain agricultural production, particularly food production. Household farming thus continues, but under drastically altered conditions.

Remittances, moreover, do not simply flow from primordial social solidarity or filial loyalty. To some extent, sons remain dependent on their fathers and other relatives even as these people depend on them. Migrants return to the village in unemployment, sickness, and old age. They count on those who stay behind to oversee their village affairs and to care for their wives and children during their absences. All family members, whether farmers, workers, petty traders, or artisans, are subject to the volatile economic conditions and fluctuating cash incomes that make it impossible to satisfy their needs entirely through the market over the long term. Their participation in a common household economy enables them to survive and reproduce. In this way not only peasant-worker households but the peasant-worker class is reproduced.

The high average age of farmers at Wad al Abbas and the predominance of younger men in other occupations does not indicate that villagers are changing from peasants into proletarians in a generation. Many of these younger men will retire into farming as they age. This transition will be facilitated by the creation of new landowners, as young landless individuals inherit land from their parents. Thus, the peasant-worker position of the villagers of Wad al Abbas will be reproduced in the next generation.

The case of Wad al Abbas suggests that once peasant agricultural production becomes dependent on the capitalist economy, it is an integral part of this economy and no longer an autonomous mode of production. The conditions of household farming and the strategies pursued by households in allocating agricultural resources are largely determined by national and international economic conditions and by the household's position in relations of production outside agriculture.

The role of non-market factors in the allocation of labor is thus exaggerated in studies that overlook the ways in which the market defines the scope and operation of non-market factors such as unpaid family labor. Furthermore, just as the labor market may determine the allocation of non-market labor, the production of subsistence crops, likewise, is related to food prices and wage levels. Therefore, subsistence production is profoundly affected by market conditions even though the crops are produced by unpaid labor and may never be marketed.

At a more macro-level, this interaction results in the transfer of value from the peasant-worker class to those who dominate the national economy. At Wad al Abbas, households are linked to the national economy both directly, through trade and wage-labor, and via the government, which controls the irrigated scheme. Value is extracted from the household in the form of cheap cotton by the state, and cheap labor and commodities by the commercial/industrial elite.

Because subsistence production at Wad al Abbas is tied to mandatory cotton cultivation, those members of the household employed outside of agriculture effectively subsidize cotton production on the scheme. This subsidy ultimately is transferred to the state in the form of cotton for which the producer price is artificially low. However, while the subsidy of farming through the household's off-farm income permits the continued production of cheap cotton, it also insures that productivity levels will remain low. Farming will not attract the most productive labor that has alternative outlets, nor will it attract capital investment. And, as cotton cultivation is not rewarded commensurately with the labor it requires, farmers will seek to minimize their labor inputs. Thus, while the scheme may continue to operate, it will do so at the very lowest levels of productivity and will receive little investment from farmers. Research on Sudan's schemes confirms this (Ebrahim 1983; Abdelkarim 1985).

On the other hand, the subsidy of household food production from off-farm income keeps the subsistence sector from total collapse. Even where subsistence agriculture depends on inputs from migrants, subsistence production and other services provided in the village "subsidize capital to the extent that they are the product of unpaid domestic labor" (Murray 1978:130). Sorghum production utilizes labor (old men and small boys) that otherwise would be idle. Women's unpaid domestic work adds value to subsistence produce and purchased consumer goods. This facilitates the household's intensification of labor. The food produced for household consumption contributes to the maintenance of the household and hence to the reproduction of the labor force. This, in turn, allows villagers to work for low wages and informal sector earnings that may fall below the actual cost of reproducing labor.

In this way, value is further extracted from the rural household as the labor power it produces sells below its cost. The Sudanese state and private employers gain cheap labor, and Saudi Arabian, multinational, and other employers abroad are guaranteed a supply of skilled and docile migrant laborers. (Saudi law prohibits emigrant workers from participating in labor organizations, and their temporary and vulnerable emigrant status mean these workers are an easily controlled labor force.)

At Wad al Abbas, the interplay between the market and household agricultural production has led to changes in the composition of household agricultural labor and in the role of farming in household economy. It has meant an intensification of household labor output and increased transfers of value from peasant-workers to other classes. At the same time, agricultural productivity levels have been depressed, and the development of agriculture inhibited.

NOTES

1. Note that throughout this discussion, the following U.S. dollar values of the Sudanese pound are used: $2.87 in 1974; $1.25 in 1981; and $1.11 in 1982.

2. It is not clear how the agricultural category was defined here since the study was conducted in urban centers.

3. In addition to the household survey, data were collected from numerous

villagers on their occupations, work experiences, income levels, and expenses among other things. This discussion draws on these various sources.

4. It is useful to estimate labor in terms of person-days so that ILO figures on the average labor requirements of cotton and sorghum on the Gezira, where agricultural production takes place under conditions comparable to those at Wad al Abbas, can be used. One day's work is counted as one person-day. If a man works 6 days out of every seven (Fridays are usually a day of rest in Wad al Abbas), in 52 weeks he works 312 days. Divided by 12 this averages 26 days/month. These person-day estimates are slightly high since social and ceremonial obligations, sickness, fatigue during fasting, marketing, and other responsibilities would probably reduce this by several days per month. Based on my observations of work patterns, I estimate a boy's labor as half a man's. For these purposes, men are 16 and older, while boys are between the ages of 10 and 15. Boys under 10 do not contribute much to agricultural work.

My man/boy ratio differs slightly from that of O'Brien (1980:420). He measures labor in terms of what he calls "agricultural labor units" with a man from the age of 17 to 64 equal to one unit and boys from 12 to 16 equal to .75. (His figures, like mine, are based on observations rather than objective measures of productivity.) I counted men 65 and over as able workers if they were still active as I did not find that farmers generally retired at this age. I counted boys between the ages of 10 and 15 as being able to do only half the work of a man. I do not know if the difference between my estimates and O'Brien's constitute real differences between our samples, nor do I find his calculations unreasonable. But young boys in Wad al Abbas are not heavily pressed to labor (though they may possess the physical capacity). They are not a disciplined work force but spend much time playing and hanging around. There is a great disparity between mens' work and boys' work in the village. Counting each boy as half a worker reflects this. Also, education through the elementary level (roughly ages 8–14) and even through the intermediate level (roughly ages 14–17) has become so widespread for boys that it must be taken as a given. Thus, in practice, most boys are not available for full-time farmwork. They help out after school and on vacations. However, unlike town schools, Wad al Abbas schools are closed during the rainy season (*khariif*), so boys are free when the planting and much of the weeding is done. The labor requirements of cultivation are not distributed evenly throughout the year but, rather, peak at certain periods and slack at others. Therefore, it is useful to estimate labor in terms of person-days per month.

5. The piecework rate for Wad al Abbas pickers in 1980–81 was 30pt/*guffa* (basket). In interviews with pickers at Wad al Abbas, they said they averaged between 3 and 6 baskets (*guffa*s) of cotton per day. Their earnings thus ranged between 90pt-£S1.80/day. Three to six baskets is much higher than the 1-1/2 *guffa*s of long-staple cotton/day that Suki pickers quoted to O'Brien (1980). And Jansen and Koch (1982) state that it is difficult for pickers to pick more than three baskets/day at Rahad (where short and medium-staple cotton are grown). I do not know the reason for these differences. At Wad al Abbas, even the smallest girl pickers of 9 or 10 were said to pick 1 or 1-1/2 *guffa*s/day. One young woman I knew picked 11 *guffa*s in one day.

Pickers are also given £S1 each time they begin picking a new tenancy and farmers pay for trucks to transport pickers from the village to the fields and back. This apparently came to about £S10 per tenancy. Furthermore, pickers from outside the village demand higher rates and some tenants reported paying up to £S1/*guffa* to get their cotton picked. All of these variations make it difficult to calculate average picking costs.

6. The negligent farmer can simply let these cotton debts pile up on his account. The credit never costs him anything since it is taken out of his cotton profits (if any) which he has already written off.

SEVEN

Off-Farm Resources, Household Farming Strategies, and Agricultural Productivity

EXPLANATIONS OF peasant agricultural behavior based on household demography and the size of landholdings are inadequate if they fail to situate the farming household in a larger political economy. Such approaches ignore the participation of peasants in relations of production outside the household, and the roles played by off-farm activities and resources in shaping household circumstances. It is often assumed that the major division in rural populations is between small and large landowners. But differentiation among farming households is not simply based in access to agricultural resources. The farming activities of most Wad al Abbas households are subsidized by income from the wage-labor and commercial work of household members. This subsidy is both direct, in the form of cash inputs to agriculture, and indirect, through the support of agricultural producers as dependent household members. The economic well-being of the household is primarily determined, not by the size of the household or its landholdings, nor by crop yields or sales, but by income from off-farm work.

Involvement in off-farm work transforms the organization and goals of peasant agricultural production. Moreover, the success or failure of the farmer may be a reflection of his access to off-farm resources and not vice versa. Variation in the production strategies of farmers arise out of differences in the positions of their households in the national and international economic system. Households with greater off-farm incomes are able to translate these into agricultural success if they choose. At Wad al Abbas, the resulting agricultural system is not a product of peasants' culture, subsistence orientation, or demography, but rather is largely a product of the interaction between the constraints imposed on household farming by conditions on the irrigation scheme on the one hand, and by conditions in labor and commodities markets on the other.

Research elsewhere suggests that Wad al Abbas is not an isolated case. The Malaysian farmers Kahn (1981:558) studied have not experienced the "differentiation into landless laborers and capitalist farmers frequently assumed to be the result of market penetration." However, the commoditization of inputs has made land and household labor less important determinants of agricultural success than they once were. Recent studies in south Asia (Desai et al. 1984) reveal that land ownership and tenure relationships are no longer the key determinants of agricultural production systems and productivity. Based on his study of Kerala, Herring (1984:203) concludes that "in a modernizing agrarian sector, land control becomes less determinative of both political and economic power." Non-agricultural incomes are vital to farming in the Nigerian and Indian regions Hill (1982) compares, where off-work both exacerbates shortages of family agricultural labor and provides cash for hired labor and other inputs to farming. Not only can cash substitute for household labor, for example, but there are inputs for which household labor is no substitute (Hill 1982:168). In Central Mexico, off-farm work affects production strategies as farmers involved in wage-labor employ more hired labor on their farms than do other peasants (Rothstein 1983). In some areas, labor migration contributes to rural class differentiation as the more successful migrants come from better-off peasants and their remittances, in turn, stimulate domestic production (Standing 1981:195–196). In a rainfed development project south of Wad al Abbas in the Blue Nile Province, off-farm income emerged as a key factor in agricultural behavior. A common characteristic of the families who opted to grow improved varieties of sorghum under partially mechanized production was a "permanent or semi-permanent non-agricultural income" (El Medani 1986:247). Families lacking such incomes could not afford to take the same risks.

VARIATION IN AGRICULTURAL STRATEGIES

A basic characteristic of agriculture at Wad al Abbas is that certain production decisions are not made by farmers, but set by state policy on the irrigation scheme. One third of a household's holdings must be left fallow, cotton production is mandatory on another third of the holding, and most inputs to cotton, beyond labor, under the control of management as are the timing and amount of irrigation to all crops. Access to irrigated land is limited, so households cannot easily bring more land under cultivation. Despite these restrictions, farmers exercise control over several key aspects of production. Thus, although subject to the same general conditions on the scheme, households' farming strategies vary. The major areas of variation in the agricultural strategies of Wad al Abbas households are as follows: *use of land*—households may keep their land or give, lend, rent (illegally), or sharecrop (legally) all or part of it to other households;[1] *allocation of labor to agriculture*—households may allocate more or less of their labor to farming;[2] and, *use of hired labor*—households may employ any amount of hired labor for cultivating cotton or other crops.[3] Other variation includes: *leaving land fallow*—farmers must cultivate the cotton portion of their tenancies fully, but are not required to cultivate the optional crop plot;[4] *increasing land cultivated*—households may increase the amount of land they cultivate beyond what they own by renting, sharecropping, or (legally or illegally) bringing land under cultivation on the scheme borders through diverting irrigation water;[5] and, *crop diversification*—households may cultivate other crops on the optional plot, instead of, or in addition to, sorghum.[6] All farmers hire tractors for furrowing sorghum land, and carry out the other operations manually. They use no fertilizers, pesticides, or other such inputs in sorghum production.

OFF-FARM RESOURCES AND HOUSEHOLD AGRICULTURAL PRODUCTION

The general conditions in agriculture described in preceding chapters affect all Wad al Abbas farmers. But farmers are able to respond to these conditions in different ways with various consequences depending on their household economies. Moreover, agriculture plays different roles in the economies of different households. For example, those who can securely afford to buy food do not have to be guided by subsistence concerns in agricultural production. Households with lucrative economic activities and investment options outside agriculture therefore are more likely to give, rent, or sharecrop their land to relatives or other fellow villagers, when cultivation is unprofitable or less appealing than the alternatives. On the other hand, wealthier households that choose to farm their land are likely to attempt to make their farms profitable, through investing in hired labor, bringing more land under cultivation, cultivating cash crops (such as coriander and peanuts) in addition to, or instead of, sorghum, and/or producing sorghum as a commercial crop. Households with sufficient off-farm income can risk complete dependence on the market for their grain consumption, and are thus able to withdraw from farming altogether or respond to fluctuation in the prices of various crops by growing other crops on the "sorghum" plot.

Most Wad al Abbas households cannot afford such strategies. As one farmer explains, "We need the sorghum. We have all these small children, and sorghum has become expensive. So, we plant sorghum and not coriander." Another farmer says, "Everybody planted grain this year because it became expensive and they need to eat." "We grow sorghum because it is expensive [to buy]," states another villager.

The area of labor is where farmers have the most leeway in decision-making. As Sørbø comments on the Khashm al Girba Scheme: "If a price fall takes place, the tenant cannot shift some of his land away from cotton. He can only vary the effort of labour he puts into his cotton production, but only to some degree due to strict supervision" (1977:141). Farmers' control over the labor process and its consequences for cotton production are recognized by scheme management. Not only are cotton operations inspected and cotton credit payments withheld until the completion of each operation, but management reserves the right to hire laborers to do the work of a farmer it deems negligent and to charge the cost (at higher than customary rates) to the farmer's cotton account. And, there is the ultimate (though more rarely invoked) sanction of taking the tenancy away.

Labor intensification represents the main strategy open to farmers for improving yields. This is also the case on the Gezira and Suki schemes where, "[t]he only means open to the tenant for increasing yields is through increasing inputs of labor" (O'Brien 1980:279). However, the uncertainty and poor returns of agriculture insure that no household invests all its labor in farming; in order to survive, households must combine farming with off-farm work. Because much household labor is committed to off-farm work, any intensification of agricultural labor (or expansion of the area under cultivation) is likely to require hired labor.

Most Wad al Abbas households lack sufficient labor to cultivate their holdings even if all their labor (subject to the division of labor by sex) were allocated to farming, which it is not. Aside from the three months of cotton picking, when labor requirements peak, 70 percent of the sample farming units[7] (26 out of 37) would be short of labor at some point in the agricultural cycle, even if they allocated all their labor to farming. Thirty percent of the farming units (11 of 37) would be short of labor for at least three months of the year in addition to the cotton harvest (Bernal 1990). This means that even if households could afford to withdraw family labor from off-farm work, which they cannot, hired labor would still be required to achieve optimal inputs of labor to farming.

Under present conditions, farming does not regularly generate profits that can be reinvested in production, therefore only households with sufficient off-farm resources can afford to invest in strategies that might lead to higher productivity. The accounting system of the scheme and the different labor requirements of cotton and sorghum are such that sorghum offers higher returns to farmers than cotton. For these reasons, and because sorghum is their staple food, virtually all farmers are motivated to achieve high sorghum yields. But farmers have different resources to apply toward this end. Variation in household farming strategies is thus particularly evident in sorghum production where households with greater off-farm incomes can hire labor as needed, and cultivate larger holdings adequately.

Households with meager off-farm incomes cannot respond to conditions in the same way as wealthier households. They neither have the means to invest in agriculture to increase returns, nor can they afford to purchase all their grain at market prices. Poor households need to produce sorghum to keep their food expenditures low. Fluctuating grain prices and inflation make it risky to gamble on buying sorghum with a scanty or unreliable income from non-farm work or with the proceeds from some other crop. Many households thus cannot afford to abandon farming in spite of low returns and high risk, because their off-farm income is too little or too insecure to support them. They thus struggle to farm even under less than optimal conditions—with inadequate labor, land, and/or cash resources—because they lack other sources of income or outlets for labor. Their lack of cash, furthermore, keeps them from pursuing strategies that might improve their yields, such as increased inputs of hired labor. As the wife of a poor farmer states, "Hired labor is for those who have wealth." Poor households, most dependent on the food crop for subsistence, are thus least equipped to be successful sorghum producers.

Differences among households in terms of their positions in the regional economy and in their off-farm resources thus affect the role of agriculture in their economies, the options open to them in agriculture, the farming strategies they pursue, and their chances of agricultural success.

The nature of these economic relationships at Wad al Abbas gradually emerged from conversations with many villagers, close, daily contact with a small number of households, and observations of conditions and activities in the village over two years. It was through this qualitative experience of the forces that shape villagers' lives, and through a vast amount of information gathered spontaneously as events occurred, that the complex relationships between agricultural strategies and off-farm employment became apparent.

Survey data[8] on the 1980–81 farming strategies and yields of the 43 sample households (37 farming units) make it possible to supplement the conclusions drawn from the qualitative component of this study with the results of quantitative analyses. Multiple regression analyses were conducted to measure the relative strengths of off-farm wealth, household labor resources, and the size of landholdings as determinants of household farming strategies. Labor resources and farm size were included as alternative independent variables because they often are seen as the primary determinants of peasant farming behavior. The inclusion of these variables also controls for a significant positive correlation between off-farm wealth and the size of landholdings. A detailed presentation of the quantitative analyses is in the appendix.

The results of the quantitative analysis support the relationships indicated by the qualitative data, and reveal off-farm wealth to be a more powerful determinant of household farming strategies than either household labor resources or farm size. While the quantitative analyses are based on the 1980–81 agricultural year, local agricultural conditions and the macroeconomic and political situation in Sudan have perpetuated these general relationships into the 1990s.

Off-Farm Wealth and Household Use of Land

Table 7.1 shows the breakdown of the sample households in terms of their use of land. Giving away or lending land represents the least degree of use of land, while retaining all holdings represents the greatest. Wealthier[9] households have other avenues of investment besides agriculture and they can more easily afford to substitute purchased grain for what they could grow. Under present conditions, agriculture is not a very attractive investment at Wad al Abbas. We would therefore expect that the greater a household's off-farm resources, the less likely it will be to retain its land for household farming; conversely, households with fewer off-farm resources will be more likely to keep their land for household farming. This is the case.

TABLE 7.1. Degree of Household Use of Land

Strategy	Farming Units	
	Number	Percent
Gave away or lent	4	11
Sharecropped all holdings	6	16
Sharecropped some holdings	2	5
Retained all holdings	25	68
	N = 37	100

NOTE: None of the sample households rented their land.

Off-farm wealth is the only variable significantly correlated with the degree of household land use (in a negative relationship, as expected); and is also the only significant predictor of household land use in regression equations controlling for the size of landholdings and household labor resources (appendix table A3). Off-farm resources, not the size of the household's labor pool or its land, determine household land use.

The fact that ten farming units relinquished control over production on their land reduces the size of the sample, to which the remaining analyses regarding farming strategy are applicable, to 27 farming units. Moreover, poor households outnumber wealthy households in the random sample to begin with and, of the 9 wealthiest households, one gave its land away and four had their land sharecropped. Thus only 4 (44.4%) of the households with the greatest off-farm wealth kept their land under household control, making it more difficult to demonstrate the relationship between off-farm resources and farming strategies.[10]

Off-Farm Wealth and Household Allocation of Labor to Agriculture

Because wealthier households derive their income from off-farm activities that consume labor resources, and because they are more able to substitute hired labor for household labor in farming, we would expect that the greater the household's off-farm wealth, the smaller will be the portion of its household labor resources allocated to farming. And this is the case.

Farming units that retained control over production on their land allocated a mean of 48 percent of their potential agricultural labor resources to farming; the range was from 0 to 100 percent, for the 26 out of

27 farming units for which this could be calculated. Potential labor resources (estimated in person-days/month) are defined as the labor of household members that could be allocated to farming within the local division of labor by age and sex. Actual labor resources represent the labor of those household members who regularly took part in farming during the survey year. Table 7.2 shows the distribution of potential agricultural labor resources among the sample households. Table 7.3 shows the actual allocation of labor resources to agriculture by these households.

Multiple regression analyses indicate that households with greater off-farm wealth tend to allocate a smaller proportion of their labor resources to farming (appendix table A.4). Neither the size of the household's potential agricultural labor pool nor the number of *feddans* it cultivated are significant predictors of the proportion of household labor allocated to farming. (However, an equation with off-farm wealth and the number of *feddans* cultivated yielded no significant results, most likely due to the positive relationship between off-farm wealth and farm size.)

Cotton, sorghum, and any other crops are cultivated over the same period, competing for household labor resources. It thus makes sense to consider the ratio of a household's labor resources to the amount of land it cultivates. However, this variable (potential household labor/*feddan*

TABLE 7.2. Potential Household Agricultural Labor Resources

Person-Days/Month	Number of Farming Units	Percent
0	2	7.4
13	0	0.0
26	2	7.4
39	2	7.4
52	8	29.7
65	2	7.4
78	3	11.1
91	3	11.1
104	2	7.4
117	0	0.0
130	1	3.7
182	1	3.7
248	1	3.7
	N = 27	100.0%

1 man = 26 person-days/month; 1 boy = 13 person-days/month.

TABLE 7.3. Actual Household Labor Resources Allocated to Farming

Person-Days/Month	Number of Farming Units	Percent
0	6	22.2
13	0	0.0
26	13	48.2
39	2	7.4
52	2	7.4
65	1	3.7
78	2	7.4
91	0	0.0
104	1	3.7
	N = 27	100.0%

1 man = 26 person-days/month; 1 boy = 13 person-days/month.

cultivated) is not a significant predictor of the proportion of household labor resources allocated to farming, either, when controlling for the effect of off-farm wealth.

Off-Farm Wealth and Use of Hired Labor in Cotton Production

Households with greater off-farm resources can afford to hire more labor than poorer households, and are more likely to allocate household labor to non-agricultural ventures which yield higher returns. At the same time, wealthier households that choose to farm are more likely to attempt to make a profit from cotton production by farming intensively. For all of these reasons, we would expect that the greater a household's off-farm resources, the more it will spend on hired labor for cotton cultivation. The quantitative data corroborate this.

The mean expenditure on hired labor for cotton cultivation is £S8.55/*feddan* (US$10.69), not counting picking, among the 21 sample farming units for whom the cash figure is available. Expenditures range from 0 to £S24/*feddan* (US$30).

Off-farm wealth is a significant predictor of household expenditures on hired labor for cotton, even when controlling for the area of cotton cultivated (appendix table A.5). Potential household labor resources/*feddan* and actual labor resources/*feddan* are also significant predictors of hired labor expenditures (in negative relationships), but have less impact than off-farm wealth. Thus, households with greater off-farm resources spend more on hired labor per *feddan* than poorer households, and it is their off-farm wealth rather than the size of their landholdings or labor forces that is primarily responsible for this.

Off-Farm Wealth and Use of Hired Labor in Sorghum Production

For the same reasons that households with greater off-farm resources tend to hire more labor for cultivating cotton, they also are likely to hire more labor for cultivating sorghum. Moreover, because the sorghum crop belongs entirely to the household, there are greater incentives for investment in sorghum than in cotton, so we would expect the relationship between off-farm wealth and hired labor to be stronger in sorghum production than in cotton production. This is the case.

On average, household expenditures on hired labor are much lower for sorghum cultivation than for cotton because cotton has higher labor requirements, and because cotton operations are inspected and negligence can be punished, while sorghum cultivation is completely at the discretion of the farmer. Fifteen out of 19 (79%) units in the sample that planted sorghum and provided data on hired labor expenses, used no hired labor at all, up to the harvest. The mean expenditure for hired labor was £S1.52 (US$1.90) per *feddan*. Expenditures range from 0 to £S11.50 (US$14.38) per *feddan*. (As with cotton, these expenses *exclude* the costs of hired labor for harvesting and threshing as these costs are dependent on yield.)

Table 7.4 shows the relationship between off-farm wealth and hired labor expenditures for cultivating sorghum. Four of the 5 households (80%) with wealth scores of 5 or higher spent something on hired labor; none of the households with wealth scores below 5 spent anything. In fact, households with greater off-farm wealth tend to hire more labor per

TABLE 7.4. Off-Farm Wealth and Hired Labor Expenses per *Feddan* Sorghum

Off-Farm Wealth Score	Farming Units' Expenses per Feddan						N
	0		less than £S10		£S10 or more		
	# units	%	# units	%	# units	%	
2–4	14	100	0	0	0	0	14
5–8	1	20	2	40	2	40	5

feddan of sorghum, even when controlling for the larger areas they cultivate and for the ratio of household labor resources to the area cultivated (potential household labor/*feddan*) (appendix table A.6).

However, it is useful to consider not only a household's potential labor resources, but the size of the labor force it actually allocates to farming and the ratio of these labor resources to the area cultivated (actual household labor and actual household labor/*feddan*). Both of these variables also are significant predictors of hired labor expenses; households with larger labor forces allocated to farming or with higher labor/land ratios tend to spend less on hired labor (appendix table A.6). Nonetheless, off-farm wealth is the most powerful of all the variables in predicting hired labor expenditures for sorghum production.

Off-farm wealth is thus a significant predictor of hired labor expenditures for both cotton and sorghum cultivation. Moreover, the positive relationship between off-farm wealth and hired labor is of greater magnitude in sorghum production than in cotton production. This is consistent with the expectation that off-farm wealth would have a greater impact on household strategies of sorghum production.

Clearly, a combination of factors shape the behavior of farming households. Household labor resources, land holdings, and off-farm resources are all important. Other factors, not systematically considered here, also affect household farming strategies—quality of land, availability of irrigation water, timing and amount of rain, the distance of fields from home, and crop prices, among others. Farmers talk about all these things when discussing village agricultural conditions in general and their own particular choices.[11] This analysis reveals the powerful impact that off-farm work and resources have on household farming, in spite of other factors such as these which are more generally recognized. Overall, differences in households' off-farm resources account for much of the variation in household agricultural strategies at Wad al Abbas.

Peasant farming cannot be understood simply as a function of household demography or farm size. The household's position in the regional economy and its members' participation in extra-domestic relations of production profoundly affect household farming. On the Blue Nile Schemes, farmers have comparatively few production options. Variation is primarily evident in the degree of household involvement in managing agricultural production and in the amounts of family and hired labor used in cultivation. Under other conditions, where farmers have more control over the production process, differences in the off-farm work and resources of households would likely result in even greater variation in their farming strategies (see Reinhardt 1988, for example.)

FARMING STRATEGIES AND PRODUCTIVITY

The ability of households with greater off-farm incomes to pay for hired labor gives them a productive advantage over households that must rely largely on unpaid family labor. If by employing labor, households are better able to meet optimal production conditions, we would expect greater expenditures on hired labor to be associated with higher yields. However, given the fact that cotton production is mandatory and farmers lack control over inputs and the costs at which they are supplied, even households that can afford to cultivate cotton intensively with hired labor may choose not to. They may instead use hired labor simply to *replace* household labor in doing the minimum required by the scheme.

While farmers all express a desire for cotton profits, few have confidence that they can attain them. Many are discouraged by the high cost of scheme-supplied inputs, particularly aerial pesticide spraying, and by the unreliability of scheme-supplied irrigation water. In addition, some farmers have come to believe that scheme management will not pay them for cotton, regardless of how much they produce. Thus, most farmers, whether using paid or unpaid labor, pursue strategies aimed primarily at holding onto their land for the sorghum plot, keeping absolute inputs of labor to cotton near the minimum required by scheme inspectors.

The sorghum crop, in contrast, not only belongs solely to the farmer and helps meet household consumption needs, but provides fodder for livestock and is a profitable cash crop as well.[12] Sorghum is by far the primary agricultural concern of village households, and farmers rank every other crop above cotton. Such priorities have been documented on Sudan's other irrigated schemes (Mohamed 1984). Households are therefore likely to use hired labor to intensify sorghum production, either by supplementing household labor or through large inputs of hired labor alone. The productive advantage gained by households that can invest off-farm income in agriculture is thus especially important in food production where incentives to improve yields are greatest.

Assessing the impact of hired labor on yields is complicated by a number of factors. First, risks are not evenly distributed among farmers — in a given year, some suffer irrigation problems or pest damage while others do not. Variation in yield reflects these factors as well as differences in labor inputs. Furthermore, while cash expenditures provide a good comparative measure of household inputs of hired labor, there is no equivalent measure of unpaid household labor inputs. The size of the household agricultural labor force does indicate how diligently that labor force worked, or how household labor was divided between cotton and sorghum. Strategies of intensive farming based on the use of household labor therefore may not be revealed here. However, given that household labor forces are generally inadequate to meet the requirements of scheme agriculture, inputs of hired labor are essential to intensive farming for most households. Therefore, we would expect a positive relationship between hired labor and sorghum yields despite the effects of unpaid labor.

Hired Labor and Cotton Yields

If many households that hire labor simply use it as a substitute for family labor, rather than to intensify cotton production, greater expenditures on hired labor will not be strongly associated with higher cotton yields. And this is the case.

Twenty-six farming units in the sample cultivated cotton and twenty-two of these provided data on yields. The mean yield is 1.48 *kantar/feddan* (88kg/ha), and yields range from a low of .40 *kantar/feddan* (24kg/ha) to a high of 5.20 *kantar/feddan* (309kg/ha). Average yields are so low that most farmers, even those with relatively higher yields, fail to meet the *minimum* requirements set by the scheme to cover production costs. Therefore, to the degree that a household does obtain better yields, these are unlikely to be of much benefit, except in minimizing debts on their cotton account and insuring that their land will not be taken away for negligence.

The number of *feddan*s of cotton cultivated is the only significant predictor of cotton yields; those who cultivate more cotton get better yields per *feddan* (appendix table A.7). Greater inputs of hired labor are not associated with higher yields, suggesting that those who hire labor for cotton cultivation are not farming more intensively but are simply substituting paid for unpaid labor. It is also possible that the high risks of cotton production reduce the positive effects of higher labor inputs. The results indicate that allocating a larger unpaid household labor force to farming does not result in significantly better cotton yields, either.

The fact that the number of *feddan*s of cotton cultivated has the strongest relationship to yields is evidence of risk factors. A Wad al Abbas household's holdings beyond 5 *feddan*s of cotton in a standard tenancy are not usually contiguous. The advantage associated with more land under cotton, thus, is largely due to the dispersal of household holdings which increases the chances that some part of the household's cotton crop might escape damage, disease, and/or water shortage.

It is significant, however, that because of the positive relationship between yields and the number of *feddans* of cotton cultivated, households with greater off-farm wealth have an advantage in cotton production, regardless of production strategy, because of their larger holdings. Moreover, off-farm resources allow them to free more family labor for better-paying off-farm work (without risking eviction), by hiring labor to cultivate their cotton.

Hired Labor and Sorghum Yields

Because there is great incentive for all farmers to achieve high sorghum yields, households are likely to use hired labor to intensify production. Greater expenditures for hired labor thus should be strongly associated with better sorghum yields, and they are.

Twenty-two of the 25 units that cultivated sorghum provided data on yields. The mean yield is 1.27 sacks/*feddan* (51 kg/ha), and yields range from 0 to 6 sacks/*feddan* (238 kg/ha). Eight units (36.6%) experienced crop failure.

Hired labor expenditures are the only significant predictor of sorghum yields in regression equations controlling for the effects of other variables (appendix table A.8). Hired labor expenditures in combination with these variables account for about 60 percent of the variation in sorghum yields.

The expectation that greater inputs of hired labor represent a strategy of intensive farming that results in better sorghum yields is strongly supported by these data. In light of the positive relationship found between off-farm wealth and expenditures on hired labor, this shows that households with greater off-farm wealth can (and do) translate it into agricultural success.

The variables included in this study are much stronger predictors of sorghum yields than cotton yields. They explain close to 60 percent of the variation in sorghum yields while accounting for less than 20 percent of the variation in cotton yields (as measured by r^2). This suggests that there are greater differences in household farming strategies on sorghum than on cotton, and/or that household strategies have greater effects on sorghum yields than on cotton yields. The fact that cotton yields, but *not* sorghum yields, are significantly related to the area cultivated may be an indication of the stronger effects of household farming strategies on sorghum yields. There is, moreover, no correlation between a household's sorghum yields and its cotton yields. These findings are consistent with the conclusion that, while most farmers pursue a strategy of minimal labor inputs to cotton production (whether primarily through household labor or hired labor), their strategies vary more widely on the sorghum crop where differences in the off-farm resources of households affect their ability to achieve better yields.

WAD AL ABBAS households with greater off-farm resources pursue different strategies on their farms than poorer households. Wealthier households tend to own more land. They are also more likely to give away or sharecrop their land. At the same time, the wealthier households that keep their land under household control tend to have large holdings. There is an even greater positive relationship between off-farm wealth and the size of land under cultivation than between off-farm wealth and land owned (appendix table A.2). This suggests that wealthier households only cultivate under favorable conditions—we have just seen that larger holdings are advantageous in terms of cotton yields. Households with greater off-farm wealth tend to allocate a smaller proportion of their labor resources to agriculture, and spend more per *feddan* on hired labor than do poorer households. Wealthier households not only hire more labor for sorghum production, but hiring labor enables them to achieve higher yields. Sorghum yields give the clearest picture of which farmers benefit most from their land, since even relatively better cotton yields rarely result in profits to farmers.

The association of hired labor with higher sorghum yields reveals that poor households, most dependent on agriculture for survival, are the least able to farm successfully, because they are unable to pay for labor. Households with little off-farm income, that can ill afford to purchase food, have the lowest grain yields. Wealthier households, not particularly dependent on the sorghum crop for food, can afford to take risks—investing in agriculture or not, and expecting some good and some bad years. The consequences of a poor harvest are not as severe for households cushioned by ample off-farm incomes; they are able to make up food production shortfalls through purchases. It is the households with small off-farm incomes that are most devastated by a poor harvest and least equipped to minimize the chances of this.

In the case of cotton, it is not possible to attribute the performance of wealthier households to their production strategies. Larger land holdings appear to be more advantageous in terms of productivity than inputs of hired labor. Part of the advantage wealthier households have thus seems to come from larger holdings. However, there is a paradox here. Few households, if any, have the household labor resources to cultivate large holdings successfully *without* hired labor; only households with access to sufficient off-farm income can afford to hire the necessary labor.

Although land is unequally distributed, to look at Wad al Abbas farmers in terms of larger versus smaller landholders would miss the point. Land alone is not enough to convey agricultural advantage. Once other inputs to production are commodities, differences in the productive capacities of households are much greater than those arising simply from differences in the size of land holdings. Furthermore, once labor markets operate, household demography loses much of its impact on farmer behavior. It is no longer the basis of agricultural production, since labor can be hired. Moreover, income from the sale of household labor can be used to pay for agricultural inputs and to purchase food.

At Wad al Abbas there may be a winning combination of off-farm resources, larger landholdings, and inputs of hired labor. At the other extreme is the poor farmer who is at a disadvantage with a small holding and who could not afford to farm a large one. It is important to remember that households fitting the latter description are much more common at Wad al Abbas than those who possess the combined advantages of ample off-farm incomes and substantial landholdings. If households with significant off-farm incomes are more likely to succeed in farming, then many households face dire circumstances on their village farms, given the realities of their position in the regional labor market.

NOTES

1. Tenancies can be legally transferred at the request of the tenant, but no payment is supposed to occur. Tenancies are exchanged for money, however. The market value of a tenancy at Wad al Abbas in the 1980–1982 period was low. A desirable tenancy, in terms of distance from the village and access to irrigation water, could be purchased for about £S100.

Rental of tenancies is not very common at Wad al Abbas, but does occur. Rental, like sale, is prohibited by scheme management. It is unlikely that this prohibition is the reason for its limited practice, however. Rather, the high risk and poor returns associated with scheme agriculture make rent unattractive to those who want land. A farmer could easily be out the rent and have nothing to show for it at all. Rental offers little profit to the landowner as rents must remain low to attract a client. A tenancy could be rented for about £S25/year between 1980 and 1982. Renting the tenancy out primarily secures access to the land for the future with little investment of any sort from the landowner.

Households may sub-contract all or part of their tenancies to others. This is strictly between the parties to the agreement themselves and in no way administered by the scheme, but is not forbidden by scheme management. Sharecropping is locally referred to as *masak bʿil nuss* (holding by half). This refers to the respective shares of landowner and sharecropper who each get half the crop. Basically there are two types of sharecropping arrangements at Wad al Abbas. In the most common, the sharecropper is responsible for all cultivation on the tenancy. All the sorghum or other optional crops belong to the sharecropper, but any cotton profits must be divided with the owner. Under this agreement, the sharecropper rarely receives any money or other help from the owner and sometimes the sharecropper must even split the cash credit for cotton production with the owner. I suspect this emerged as an adjustment to the lack of cotton profits in the past ten years or more when cotton credit came to be seen as the only cash income most farmers can expect from their tenancies. In the absence of cotton profits, if the sharecropper kept all the credit payments, the owner would receive nothing for the use of his land.

In the second type of sharecropping arrangement, the sharecropper receives cash from the owner to help meet the expenses of cultivation and must then split the sorghum harvest with him. The sharecropper keeps the cotton credit and any cotton profits.

Sharecropping is not a profit-oriented strategy for the land-owning household. It is simply a holding operation whereby the right to the tenancy is maintained in spite of the household's inability or unwillingness to perform the obligatory cotton cultivation. Because of low yields and risk in sorghum cultivation, even in the second type of sharecropping, landowners do not stand to gain much.

Sharecropping at Wad al Abbas requires very little involvement on the part of the landowner who has no part in the production decisions of the sharecropper. Except for owners who received part of the sorghum crop, farmers who had sharecropped their tenancies did not even know what the sharecroppers' yields were. Sharecropping represents, after gift, sale, or rental, a method of little involvement in terms of cash, labor, or supervision on the part of the tenant household.

2. The intensity of household farm work and the number of household members who participate in farming varies. On the cotton crop, however, tenants cannot reduce labor inputs below a certain point without risking eviction. However, they may employ hired workers to do the work.

3. Even under optimal conditions, many tenants would be short of unpaid labor. Thus, the use of hired labor is virtually a prerequisite of intensive farming on the scheme. Low hired labor inputs may be due to lack of funds for hiring labor or to conscious disinvestment in agriculture in favor of other endeavors. In the case of cotton, low hired labor inputs may even represent "passive resistance" to conditions on the scheme, as some have argued.

Cotton picking, the pulling out of cotton plants after the last picking, and sorghum harvesting are the operations where hired labor is most heavily relied on. Most households resort to hired labor in these operations. Households that use more hired labor, use it not only for these operations but for sowing and weeding cotton and/or sorghum. The reverse was not found—a household using hired labor for sowing sorghum but not for harvesting it, for example. It makes sense that households trying to conserve on hired labor would resort to it only for the most labor-intensive or heaviest tasks while carrying out the other tasks themselves.

4. There is no advantage to the household in leaving land fallow. Households do this because of last-minute shortages of cash or labor and failure to arrange with a sharecropper, or because the household does not want to give up all the optional plot to a sharecropper yet lacks the labor or cash to cultivate it fully themselves.

5. In the first two decades of the scheme, additional tenancies were available on request for sons. This is no longer the case, although tenancies occasionally become available when management revokes the landrights of cultivators it deems negligent. Such land is allocated to new owners at no cost. Conditions on the scheme are less than a free market in land, but tenancies are bought, sold, and rented illegally. To temporarily increase their holdings without paying cash, some tenants sharecrop. And they can (with permission in some cases, illegally in others) bring land on the edge of the scheme under cultivation by diverting water from irrigation canals. (Barnett, 1977, also refers to this practice on the Gezira Scheme.) The expansion of land under cultivation represents an investment in agriculture, entailing increased labor requirements for the tenant household, either in unpaid household labor or hired labor.

6. Other crops grown on the scheme besides sorghum are wheat, coriander, and peanuts. Coriander and peanuts are purely cash crops destined for sale. Wheat is not a local subsistence crop but is consumed as a luxury food as well as sold. Of course, sorghum is sold, too, and grain prices were high and rising during the 1980s. In addition to differences in market price and consumption value, crops have various labor and water requirements which farmers take into account when considering their production strategies each year.

Sorghum was by far the most common optional crop grown during fieldwork. The great majority, if not all, of the producers of sorghum grew it mainly for household (and sometimes additionally, livestock) consumption. As a rule of thumb, the sale of sorghum was only considered in cases of desperate cash needs or if grain supplies clearly exceeded maximum anticipated consumption.

7. As noted in chapter 4, because of local practices of polygyny and uxorilocal residence, the household and the farming unit are not always identical (although in most cases they are). Because there were multiple households of polygynous men, the 44 households in the sample constitute 38 farming units. One household did not provide detailed information, reducing the sample size to 43 households or 37 farming units.

8. Household heads were asked about landholdings and tenure relationships, such as rental or sharecropping, which crops they had planted, how much land was under each crop, and the yields and proceeds (if any) from each crop, for 1980–81 and 1981–82 (see Appendix for a discussion of 1981–82 data). They were also asked how much they spent on hired labor for each crop they cultivated and on which operations they used hired labor, as well as which family members participated in production. Leaving land fallow, increasing land cultivated, and crop diversification had to be excluded from the quantitative analysis since too

, few households in the sample engaged in these practices to yield any significant results.

9. Where the multiple households of a polygynous have different off-farm wealth scores, the highest score is taken as the index of wealth for his farming unit.

10. In considering households' farming strategies and yields, households that had their land cultivated by sharecroppers were not included since little information regarding the production process was available from the land-owning households, and information on yields was also limited due to the fact that the sharecropping agreement generally provided for the splitting of cotton credit and/or profit while the sorghum crop belonged to the sharecropper. Landowners thus took no interest in sorghum yields, or even cotton yields, since in none of the cases were there any profits.

11. This study does not attempt to model the actual decision-making process in peasant agriculture. It seeks to uncover patterns in behavior and explain them with reference to a few key factors.

12. There is, then, not so much a dichotomy between cash *crops* and subsistence *crops*, as is often assumed, but between subsistence *producers* and cash crop *producers*—who may be cultivating the same crop. Even this distinction is a blurred one because poor farmers may actually sell crops they need for their own subsistence to meet other necessary expenses (Hill 1982). This is true at Wad al Abbas, as some cash-poor farmers are forced to sell part of their harvest in the fields to pay the hired laborers they used in cultivation.

EIGHT

Peasant-Workers and Development

THIS BOOK has argued that third world smallholders can best be understood as part of a peasant-worker class. Too often theories of peasant behavior have viewed agriculture in isolation from the other economic activities of rural people. This has led to a methodological focus on characteristics of the farm and farming household, such as land size, crop types, and household demographics, while neglecting conditions in labor markets and state policies that affect agriculture directly and indirectly. Research on third world workers, on the other hand, has often emphasized the differences between peasants and proletarians and therefore overlooked the role of wage employment and labor migration in transforming and sustaining farming systems.

The people of Wad al Abbas force us to recognize the significance of off-farm employment and income in peasant household economy. At Wad al Abbas, labor migration, remittances, and other aspects of proletarianization are changing the rural economy in profound ways. Agriculture on the Blue Nile Scheme is not profitable for the majority of farmers. It does not even provide people with subsistence. Neither sorghum production nor cotton profits are sufficient to maintain households from one agricultural year to the next. Peasant livelihood has been undermined by the state-imposed and regulated irrigation scheme. Farming is no longer a buffer against proletarianization or a viable alternative to it. Rather, the high risks associated with the irrigation system and farmers' loss of control over the production process, coupled with high production costs and low cotton profits, make access to cash from other sources vital to farming households. Both household agricultural production and wage-work or informal sector activities are essential to household reproduction and to the maintenance of the community. Household farming is dependent on inputs from off-farm work.

Trade and wage-labor, not agriculture, are the basis of wealth differences within the village. Economic strata are not based in differential control of agricultural resources but in accumulation of mercantile capital. Although a few villagers are accumulating wealth, the village is not dividing into capitalist farmers and a landless proletariat, nor is differentiation emerging primarily between large landowners and other peasants. Rather, involvement in off-farm work is changing households' agricultural options and strategies. The opportunities for investing household labor outside of agriculture alter its imputed value in household farming. The composition of the household farm labor force is determined by conditions in the wider labor market. Households' use of land and their use of paid and unpaid farm labor are more influenced by their access to off-farm resources than by the internal dynamics of their household economies in terms of family labor resources and land ownership. The productivity of their farms, moreover, is directly related to the ability of household members to generate income through off-farm work.

The number of Wad al Abbas villagers actually engaged in wage labor is still quite small. However, the overall process of converting a community of autonomous rural agricultural producers into a proletariat is far more advanced than the number of those actually engaged in wage-labor would suggest. The conditions of production in agriculture resulting from state control have pushed some segments of the population into non-agricultural work. While few villagers are employed as wage-workers, villagers can no longer support themselves through production on their own land. Many are forced by limited employment opportunities to work as self-employed petty traders. The harsh economic conditions outside of agriculture, in turn, compel most households to sustain agricultural production.

Households at Wad al Abbas are struggling to survive and, if possible to accumulate wealth with whatever combination of agricultural and non-agricultural strategies they are able to pursue. This means that changes in the opportunity structure of labor markets and the commercial economy as well as changes in agricultural conditions affect their farming strategies. Where a household is involved solely in agriculture, the goals of household reproduction and prosperity are synonymous with successful farming. But where the household depends on diverse economic activities, household resources may be diverted from agriculture to other pursuits, and income from these pursuits may be vital to household farming. At Wad al Abbas, income from wage-labor and trade are used to pay for hired farm labor and to support dependent household members who form the unpaid household farm labor force.

For a few households, non-agricultural activities are so profitable that they abandon agriculture altogether. For other households, off-farm income allows them to invest in farming, supplementing or replacing household labor with paid labor, for example. For the majority who are less fortunate, the need for off-farm income and the associated allocation of household labor to off-farm work result in a precarious economic cycle. We have seen that poorer Wad al Abbas households, most subject to shortages of cash for inputs to agriculture, are at the same time most likely to keep and cultivate their land. Their agricultural performance suffers because their off-farm earnings are insufficient to pay for inputs and their off-farm work leaves them short of household farm labor. At the same time, because the off-farm income of these households is so limited and insecure, food production remains vital to household welfare; so they must continue to farm even with inadequate resources. Significantly, the disadvantage suffered by such households is not as evident in cotton production, which receives most attention from scheme management and researchers, as it is in food production. Thus, it has gone largely unnoticed.

Beyond the level of the household, this process has negative implications for agricultural development and productivity. At Wad al Abbas, the low returns from agriculture have led the most able and productive villagers to leave farming; household agricultural labor forces and the population of farmers are made up of old men and young boys. Moreover, the low returns from agriculture, coupled with the need for hired labor, limit the productivity of households with meagre off-farm incomes. At the same time, even for households with off-farm wealth to invest in agriculture, the returns are relatively low and risky. Wealthier households therefore tend to disinvest, transferring their land to poorer households. Thus, the overall pool of farming households may be made up more and more of the poorest households that are least able to meet optimal conditions of production. Poor households, moreover, may become progressively more impoverished as they continue to farm because they fall short of even their subsistence grain and incur debts to the scheme for cotton inputs.

State policy on the agricultural scheme at Wad al Abbas has contributed to the impoverishment of a portion of the village population. Conditions imposed by the scheme compel households to intensify their labor through combining agricultural production and various trade and wage-labor activities. This results in greater expropriation of value from them by the government through cotton production on the scheme, where returns to producers are scant.

To the extent that subsistence production on the scheme contributes to the reproduction of labor, it provides cheap labor to the public sector and to private capital in the Sudan and abroad, through labor migrants. The poor conditions in agriculture coupled with limited opportunities for wage employment also push villagers into labor-intensive small-scale trade and craft activities. Comparatively few villagers are able to amass wealth and become big merchants. Most are self-employed petty traders who fill economic niches not profitable enough to attract capital. They further the expansion of commodity relations, becoming a link in the chain of accumulation that connects "isolated" rural households to the national and international market system.

The unpaid domestic farm labor producing subsistence crops, and the communal norms of mutual aid, sharing, and support of dependents stretch wages and meagre earnings, making it possible for men to survive periodic unemployment and other harsh working conditions. The non-migrating population provides the basis for the wage and self-employment of migrants. Agriculture, wage-labor, and the informal sector are thus linked through the peasant-worker class that must combine them to survive.

The decline of agriculture and the consequent growth of off-farm employment have undermined the local basis of village economy. Because agriculture is maintained through off-farm income (to support the unpaid household labor force and to pay for hired labor), farming cannot cushion villagers from upheavals in the regional and international economies. The very conditions of household agricultural production are determined by these economies. The heavy reliance of Wad al Abbas households on purchased rather than home-grown food makes them especially vulnerable. The village is becoming largely a repository of marginal labor, old men, women, and children, a place to be sick, to visit on holiday, and to observe ceremonial obligations, but no longer a vital economic entity. This makes villagers ever more dependent on the market for their wages, salaries, and trade profits, and for purchased grain and other consumption items. They are less in control of their own destiny and more subject to the vagaries of the national and international economies over which they exercise no control.

In the case of Sudan, it is also significant that the most skilled, productive labor has not only left agriculture, it has left the country. Thus, the impoverishment and proletarianization of peasants are not so much contributing to the accumulation of capital within Sudan as outside it.

THE PEASANT-WORKER CLASS AND DEVELOPMENT THEORY

The case of Wad al Abbas is important because it challenges some of our assumptions about peasants, agricultural development, labor migration, and class formation. The working class cannot be defined on the basis of access to means of production. Nor can we assume that capitalist development in the third world is inexorably destroying peasant production and creating a landless proletarian labor force. This study points to the connections between labor migrants and their rural communities by showing that the peasant activities of non-migrants are transformed. It draws attention to the roles of off-farm work and resources in shaping household farming strategies, revealing important links between urban and rural, and between peasants and proletarians.

While the main division in third world nations sometimes appears to be that between city and country, relations of exploitation are not between urban and rural populations, but between classes. There are elites in the countryside and poor in the cities. Moreover, while the locus of their work may be different, the conditions of existence of peasants and proletarians are increasingly determined by the same factors as they are subjugated by the same capitalist elite. Labor markets, rural-urban migration, wage levels, and patterns of work and accumulation in the informal sector are thus key factors shaping rural stratification and agriculture.

The relationship between household agricultural production and the capitalist economy is not simply a matter of subsistence farming keeping wages low or peasants becoming proletarians. The conditions of household farming themselves are altered and there is no clear line between peasants and proletarians who both depend on rural non-market production and income from wage or informal sector employment. It is particularly the involvement in diverse relations of production and exchange within the same household economy that facilitates transfers of value to capitalists and the state from peasants and proletarians alike. Many factors contribute to this process; the irrigation scheme at Wad al Abbas is but one example.

Communities like Wad al Abbas are not rare, and the ranks of peasant-workers are likely to grow as agricultural resources are concentrated in the hands of elites and rural people are drawn into relations of production organized by capital in and outside of agriculture. For example, in Central Peru "external sources of income are often pivotal in meeting household needs and in perpetuating peasant forms of production" (Long and Richardson 1978:202). In Malaysia, "Without wage work, it is doubtful if the rural structure based on extraction of labour tribute [from sharecroppers] can survive for long" (Halim 1983:266). Scott's (1976) treatise on peasant moral economy mentions the "labor-exporting village" that is more affected by conditions in the wage-labor market than by the prices of agricultural commodities or a poor harvest. In areas of Kenya's Central Province over 60 percent of the income of farming households is from wages (Cowen 1981).

Furthermore, while there are peasants who retain a great degree of self-sufficiency (e.g., Donham 1985) clearly they will become a smaller and smaller minority as capitalist agriculture, development projects, agribusiness, contract farming, population pressure, and ecological decline restrict the resource bases and production options of farmers. "[E]ven if there are still millions of people in the world, not directly involved in capitalist employment, how many are still capable owing to the social disruption, famine and wars it brings about, of producing their own subsistence and feeding their children?" (Meillassoux 1981:140).

IMPLICATIONS FOR AGRICULTURAL DEVELOPMENT POLICY

This study should change the way we think about two key agricultural resources—land and labor. It should also lead us to question the current focus of policy on setting the "right" prices of agricultural commodities or raising food prices as ways of stimulating agricultural productivity.

Land

As capital increases in importance in the agricultural production process, the significance of land decreases. Differences in the productive capacities of farmers, then, are not necessarily based in unequal access to land. While the distribution of land must be considered in any agricultural plan, it should be placed in a broader economic context. Rural economic strata can no longer be assumed to be land-based. Therefore emphasis on land reform, property rights, or the optimal size of holdings is inadequate to address the problems of agricultural development. Simply broadening access to land will not necessarily improve agricultural productivity or rural welfare unless coupled with access to other resources that will allow landowners to realize the productive capacity of their land and exercise control over their produce. It can alternatively lead to the proliferation of peasant-workers who can undertake poorly remunerated wage-work because they are partly supported by household farming and who can continue to farm unprofitably because they are partly supported through wage-work.

Labor

Rural populations are engaged in work outside of agriculture. The organization and labor requirements of these activities interact with those of agriculture. There may be much less elasticity in the labor supply to agriculture than is often presumed when, for example, labor migration is viewed as evidence of "underemployment" in the countryside.

Many development projects require an intensification of agricultural labor—more careful weeding, application of fertilizer, etc. When rural populations are viewed solely as peasants this extra labor is presumed to be available at no cost from family members. Recognizing the off-farm work of rural populations and the links between agriculture and other sectors of the economy reveals the tradeoffs between farming and off-farm work. Furthermore, the operation of labor markets in and outside of agriculture means that increased labor requirements of farming may have to be regarded as commercial inputs. This would alter cost/benefit calculations so that they might yield a more realistic assessment of the value of proposed agricultural intensification plans.

The non-agricultural work of rural populations can, under certain conditions, serve positive roles in enhancing agricultural productivity and rural welfare. Off-farm work offsets the risks and low incomes in agriculture as well as the seasonality of agricultural labor. Off-farm work can generate capital for investment in agriculture. Where land pressure is a problem, off-farm work may also reduce competition for land or increase the viability of smaller plots. Agricultural programs should incorporate non-agricultural features that will serve these purposes. Involvement in off-farm work is inevitable as successful farmers are able to provide education, training, and even capital for their children and as poor farmers are forced to supplement low farm incomes from other sources. Agricultural policies and income-generation projects should seek to enhance the ability of poor farmers to earn a livelihood. Access to land alone is not sufficient, nor is access to agricultural credit which may become a burden of debt. Agricultural incomes must be increased wherever possible, and viable alternative means of making a living be expanded. Small-scale rural industry or vocational training in skilled crafts such as carpentry, and in areas like mechanics, electrical work, and plumbing, that are necessary for the maintenance of rural infrastructure might be particularly appropriate.

Prices

We must also question the emphasis on prices, especially the wisdom of deregulating food prices, which assumes that urban populations benefit from low food prices at the expense of rural people and that higher prices will cause farmers to produce more food. Prices may affect the distribution of produce more than they can affect the distribution of productive resources or the organization of production. Therefore, prices alone cannot overcome organizational obstacles to development in the agricultural sector. Poor farmers thus may be unable to expand production regardless of price incentives. Moreover, as this research has shown, farming households are often net consumers of food and suffer rather than benefit from rising food prices. This is especially true of the poorer segments of the rural population.

Broad agricultural reorganization as well as changes in employment conditions and opportunities are needed. Farmers must gain greater control over productive resources and over the production process in agriculture so that they can satisfy their consumption needs and generate a surplus which they can reinvest in production. Wage levels must be improved so that the labor force can be reproduced without the subsidy provided by subsistence farming.

RURAL PROJECTS should include non-agricultural sources of income and employment as an integral feature. Creating a highly productive agriculture may reduce levels of employment in agriculture as agricultural incomes rise. Alternatives thus must be created hand in hand with agricultural development. Agricultural policies and the structure of the agricultural sector cannot be addressed separately from the other sectors of the economy. Policies must take into account the fluid boundaries between peasants and proletarians and between urban and rural populations.

Specific policies and programs must be shaped by local conditions and concerns. This can best be done through the involvement of local populations in the planning and implementation of development policies for their area. Obviously, this process will be slow and difficult because it requires building viable institutions from the ground up and creating a participatory system. But, as Richards has pointed out in the case of African agriculture, "the 'dramatic modernization' option has a track record so poor that a return to slower and more incremental approaches must now be given serious and sustained attention" (1985:160).

Leaving key decisions about the allocation of resources and the organization of production to the whims of market forces or the interests of a small capitalist class is too risky. On the other hand, the administrative capabilities of developing nations are already overtaxed. It is unwise to call for massive state management of agriculture or any other sector of the economy. Certainly, neither of these strategies have worked well in the African context. Small-scale organization and institution-building from the grass roots would appear to be the best strategy for creating the building blocks of democratic societies and promoting the rational use of resources to benefit the largest number of people.

POLICY IMPLICATIONS FOR WAD AL ABBAS AND SUDAN

One reason for the failure of current policies on Sudan's irrigated schemes is that the real conditions of the majority of the rural population are not taken into account. Their off-farm work is often seen as "absenteeism." Such an approach ignores the dynamic process of transformation these populations have experienced, a process in which the schemes themselves have played an important role. The reality is that few households, if any, could afford to continue farming on the scheme if they did not have off-farm work. Moreover, it is the households with lucrative off-farm activities that are most able to farm successfully because they can afford to supply all necessary inputs.

In terms of the future of Wad al Abbas, unless there is a major reorganization of the scheme, agriculture will not be a prime source of subsistence or accumulation for much of the population. Their only recourse will be to sell their labor or to attempt to create opportunities for self-employment in the informal sector. If present trends continue, there simply will be no way for an ever-larger group of people to be productively employed. Clearly Sudan lacks capitalist units of production capable of absorbing all of this labor. And, while Saudi Arabia and neighboring states have provided a temporary escape valve, any constriction in these economies could have severe consequences. In addition, from the national perspective, Sudan cannot afford to continue losing its most productive and skilled workers.

To obtain the necessities of survival, people will migrate in ever-greater numbers and accept whatever work they can find. The Sudan already has a highly mobile labor force (ILO 1976). Villagers from Wad al Abbas, by their own reports, endure harsh conditions and painful separations from their families because they see few economic alternatives to labor migration. The economic balance thus achieved, however, is precarious and extremely unlikely to lead to development in terms of a more optimal use of resources or distribution of wealth.

Dire conditions in the labor market and the national economy assure that the Wad al Abbas agricultural scheme will not run out of the primarily marginal labor that sustains it. At the same time, the involvement of villagers in informal sector and wage employment will continue to grow. The cycle of extracting value from the self-exploitation and intensification of labor of the peasant-worker class will continue—without development—unless conditions of production in agriculture and outside of it are changed.

The aim of any policy reforms on Sudan's irrigated schemes must be to increase agricultural productivity and to enhance the welfare of the rural population. The most essential feature of such policies would be the transfer of control over production decisions and productive resources to the farmers. Plans that depend on the ignorance and powerlessness of rural people cannot lead to development. These people are never quite as ignorant or powerless as they may seem. The people of Wad al Abbas have demonstrated this. But their household-level solutions to the problems of agricultural insecurity and rising cash needs cannot bring about the restructuring of the economy necessary for development. That requires organization and institution-building. Involving villagers in reforming the schemes and planning for their ultimate takeover thus is not only vital to improving agricultural performance, but would foster the organization of the rural populace, providing a basis for other development efforts.

Giving farmers greater control over agricultural production and supporting their efforts to produce food would stimulate agricultural productivity and raise agricultural incomes. This in turn would facilitate another important goal of agricultural policy which should be to slow the rate of rural proletarianization and labor migration so as to bring them in line with the creation of jobs outside of agriculture. Labor migration is currently disrupting the lives of families and the pattern of community life. It creates a gender gap as women are left behind in rural areas to raise families and manage household resources, while income is controlled by absent males. On a larger scale, migration exacerbates regional imbalances in development, both nationally and internationally.

To reduce labor migration, rural non-agricultural production should be developed in tandem with agriculture. The location of production in rural areas would have the further advantage of making it easier for women to gain access to sources of income. Currently, private investment in Sudan goes into trade and transport. Entrepreneurship and managerial expertise are highly developed in the commercial sector; skill and capital would appear to be available, but until trade loses what some have identified as its "privileged and protected position in the economy" (ILO 1976:452), they will not be invested in production. Government policies that make trade less attractive could shift capital into production. Given the difficulties of supplying fuel, spare parts, and other imported inputs to production in the Sudan, rural industry would likely suffer many of the same problems as the irrigated schemes. Capital and import-intensive strategies should be avoided in favor of simple technology and reliance on locally available inputs.

PEASANT-WORKERS AND CLASS FORMATION

The link between peasants and proletarians was recognized by Lenin who held that the proletariat are not necessarily landless, that, in fact, "The allotment of land to the rural worker is very often to the interests of rural employers themselves, and that is why the allotment-holding rural worker is a type to be found in all capitalist countries" (in de Janvry 1981:99). Similarly, Cohen, Gutkind, and Brazire write:

> The working class in the Third World, as normally defined is thus small. . . . However, in our views it is not necessary to adopt a restrictive definition of a "working class" in order to understand how surplus value is

expropriated in Third-World societies. The collapse of subsistence economies in the face of capitalist penetration has led to the growth of a large group of workers who are ambiguously and simultaneously "semiproletarians" and "semipeasants." (1979:12)

They go on to point out that casual workers, and petty entrepreneurs and their like "hover constantly between capitalist and non-capitalist modes of production," and add that: "In this very 'murky' sector of the economy the costs of production and reproduction of labor are absorbed in the non-capitalist sector, thus allowing capital to extract a significant portion of its surplus from the frantic efforts of the poor to survive" (1979:20). Their implicit model (capitalist vs. non-capitalist) is overly dualistic, but they have nonetheless captured the blurred distinction between peasants and proletarians that is at the heart of underdevelopment.

The concept of articulated modes of production is inadequate to represent the degree to which "traditional" relationships, "subsistence agriculture," and so on have themselves been transformed and restructured so that the opposition between capitalist and non-capitalist is not clear. There has been a blending to form a new whole. While it differs from full-fledged industrial capitalism, it cannot be described as capitalist and non-capitalist—any more than women's unpaid labor in the home under industrial capitalism represents a separate mode of production. Rather, it represents a way in which value is extracted without paying wages, but its structure and organization is largely determined by conditions in capitalist labor and commodities markets.

The ability of peasant producers to survive under capitalism through intensifying their unpaid labor and selling their products below their actual costs explains the persistence of peasant forms of production within capitalist economies. The peasant-worker household sells not only its agricultural products but its labor. In the peasant-worker household the labor-time available for "self-exploitation" on the farm is the labor of household members for which there is little or no market. Such peasant production does not have an existence independent of the labor market and can only be understood in reference to it. This, I would argue, is the case of much peasant agriculture today. It is integrated into a larger economic structure, not as a mode of production, but as a subordinate part of this larger structure. Where peasant agriculture is sustained in part through wage-labor and self-employment, peasants must be seen as part of a working class, a peasant-worker class.

The peasant-worker class, moreover, is not necessarily transitory. It is not simply the phenomenon of a generational change from peasant to proletarian. Present conditions in many areas of the world make the complete proletarianization of this class unlikely. Capitalist production is not expanding with enough speed or consistency to provide employment at living wages. In fact, processes of reverse migration and re-peasantization have been described in some areas, as crises in the labor market constrict employment. Thus, individuals and households enter and retreat from various relations of production, but the reproduction of the class through the combination of household agricultural production and wage/informal sector employment continues.[1]

A transformation of the position of the peasant-worker class and their conditions of existence would challenge other class interests and the course of development now underway. But changes could be brought about through the uniting of peasants, wage-workers, and informal sector workers. As populations are drawn more and more into the peasant-worker class, the concerns of workers and farmers converge. The diversity within this class and the shifting and unstable work experiences of its members mean that class consciousness is not likely to emerge in the way centralized industrial production creates it among workers. Nonetheless, common interests resulting from the dependence of workers on rural producers and the dependence of peasants on employment and remittances can provide a basis for mobilization. There is some evidence that this is occurring as urban workers and rural farmers alike are hard hit by IMF austerity measures, for example. The boys' riot at Wad al Abbas in 1982 was part of countrywide protests over economic conditions. Mass strikes and demonstrations in Sudan over the last decade have not centered around wage disputes which affect a minority, but around consumer prices and shortages of essentials—issues that hit the poor hardest and link diverse economic groups. IMF riots in many countries over the price of bread and other basics reflect the same pattern. Such things suggest that urban-rural, agricultural-industrial divides can be bridged by the peasant-worker class.

If economic development is brought about by a dominant capitalist class, the peasant-worker class will be transformed into a full-fledged proletarian class as a larger and larger proportion of workers will be able to be reproduced independent of peasant agriculture, and peasant agriculture will come to be replaced by capitalist agriculture. Peasant-workers then will gradually become only a small fragment of the working class.

But, in much of the third world, the converse will remain true for some time to come. Full-fledged proletarians will be the minority of a working class which cannot be reproduced independently of peasant agriculture. Low productivity peasant agriculture will be sustained through wage and self-employment, and larger numbers of peasants and pastoralists will be drawn into the ranks of the peasant-worker class. The expansion of this class will help make continued accumulation possible for elites without the transformation of production, by intensifying the labor of greater numbers of people. Global farming populations in the immediate future will thus be made up more and more of cultivating workers.

NOTES

1. The peasant-worker class may not be a unique feature of peripheral economies. Holmes and Quataert (1986:192) argue for the great continuity of "the worker peasantry, throughout European industrial development—from proto-industry to the post-world War II era." However, its prevalence and significance in underdeveloped economies is greater since it numerically dominates the working class.

APPENDIX

Quantitative Analysis of Farming Strategies and Productivity

DUE TO the close association of key variables, such as the size of landholdings and off-farm wealth, simple correlations (Pearson's r) were not sufficient to test for the predicted relationships. Correlations were performed to identify which variables were closely associated, then multiple regression analyses were conducted. Multiple regressions make it possible to assess the impact of an independent variable while controlling for other variables. They provide an indication of the relative strengths of alternative independent variables as predictors of a dependent variable. Multiple regression equations are attractive compared to analyses such as Chi Square because regression equations allow the use of continuous variables, preserving some of the complexity of the data. Moreover, because of the small sample size, there would be too few cases in some instances if discontinuous variables, such as land rich/land poor, high off-farm wealth/low off-farm wealth, large labor force/small labor force were used. Even so, the small sample size and close association of some variables made it necessary to limit the number of independent variables in each regression equation to two in order to reveal statistically significant relationships.

Table A.1 shows how major variables were coded for analysis. In correlations and regression tables all results are tested for significance at the .05 level.[1] While data were collected on 38 farming units, the size of N varies because not all households kept control of their land or planted sorghum, etc. and, on a few questions, data for some households were incomplete.

TABLE A.1. Coding of Major Variables

Independent Variables

Off-farm wealth: 2–8 from lowest to highest
Labor Resources: number of person-days/month
Land owned, land cultivated, and area of each crop cultivated: number of *feddans*

Dependent Variables

Degree of Household Use of Land:
Gift or Sale = 0
Sharecropping all household's land = 1
Sharecropping part of the household's land = 2
Retaining all household land = 3
  (There were no cases of rental in the sample.)
Proportion of Household Labor Resources Used: number or fraction of person-days/month
Hired Labor Expenditures: number of Sudanese pounds/*feddan* of the crop
Cotton Yields: number of *kantar*/*feddan*
Sorghum Yields: number of sacks/*feddan*

Since the effect of off-farm wealth on farming strategies is measured against that of alternative independent variables, it is useful to consider the relationship between off-farm wealth and these other independent variables (table A.2).

Off-farm wealth and the size of holdings are positively and signifi-

TABLE A.2. Correlations (Pearson's r) Between Off-Farm Wealth and Alternative Independent Variables

Independent Variable	Correlation with Off-Farm Wealth		
	r	s	N
Feddans owned	.345*	(.325)	37
Feddans cultivated	.517*	(.396)	25
Potential household labor	.257	(.325)	37
Actual household labor	.059	(.381)	27
Potential labor/*f* cultivated	−.034	(.396)	25
Actual labor/*f* cultivated	−.264	(.396)	25

NOTE: All tests are two-tailed.

*Significant at .05 level.

cantly correlated. Wealthier households tend to have larger holdings. The number of *feddan*s actually cultivated by a household (regardless of how they are acquired, and *not* including fallow or land owned by the household but cultivated by sharecroppers) is even more strongly correlated with wealth than is land owned. This indicates that not only do wealthier households own more land, but that the strategies they pursue increase their advantage over poorer households in terms of land size. It is important, then, to control for the effects of land size on farming strategies in assessing the explanatory strength of off-farm wealth. There are no other significant relationships between off-farm wealth and the alternative independent variables.

Tables A.3 through A.8 report the results of multiple regression equations measuring the relative strengths of different variables as predictors of household farming strategies and as predictors of yields. Land size and labor resources variables are always included as alternative independent variables whose effects are measured against those of off-farm wealth in predicting farming strategies and against those of hired labor in predicting yields. Since labor resources are measured in several different variables, Pearson's r were conducted to determine which of these to include as independent variables in particular equations, based on whether they had a significant or nearly significant correlation with the particular dependent variable. However, even where no significant correlation was found, labor resources are included as an alternative independent variable.

The relationships identified through the multiple regressions reported

TABLE A.3. Regression Equations Predicting Degree of Household Land Use: Raw Regression Coefficients and t-ratios

Independent Variable	Equation Number			
	1		2	
	coef.	t-ratio	coef.	t-ratio
Intercept	2.71	7.01	2.93	6.99
Off-farm wealth	−.21*	−2.32	−.15*	−1.74
*Feddan*s owned	.02	1.55	—	—
Potential household labor	—	—	−.0005	−.13
N	37		37	
r^2 adjusted	.100		.037	

*Significant at .05 level, one-tailed test.

TABLE A.4. Regression Equations Predicting the Proportion of Household Labor Resources Allocated to Farming: Raw Regression Coefficients and t-ratios

Independent Variable	Equation Number			
	1		2	
	coef.	t-ratio	coef.	t-ratio
Intercept	.73	5.55	.60	3.74
Off-farm wealth	−.07	−1.70	−.07*	−2.16
*Feddan*s cultivated	−.001	−.23	—	—
Potential labor/*f* cultivated	—	—	.02	1.31
N	25		25	
r^2 adjusted	.097		.160	

*Significant at .05 level, one-tailed test.

TABLE A.5. Regression Equations Predicting Cash Spent on Hired Labor per *Feddan* Cotton: Raw Regression Coefficients and t-ratios

Independent Variable	Equation Number					
	1		2		3	
	coef.	t-ratio	coef.	t-ratio	coef.	t-ratio
Intercept	−.69	−.23	4.37	1.16	5.49	1.47
Off-farm wealth	2.02*	2.33	2.13*	2.92	2.27*	3.32
*Feddan*s cotton	.26	1.09	—	—	—	—
Actual labor/*f* cultivated	—	—	−1.07*	−1.76	—	—
Potential labor/*f* cultivated	—	—	—	—	−.87*	−2.15
N	21		21		21	
r^2 adjusted	.359		.417		.456	

*Significant at .05 level, one-tailed test.

TABLE A.6. Regression Equations Predicting Cash Spent on Hired Labor per *Feddan* Sorghum: Raw Regression Coefficients and t-ratios

Independent Variable	Equation Number							
	1		2		3		4	
	coef.	t-ratio	coef.	t-ratio	coef.	t-ratio	coef.	t-ratio
Intercept	−.28	−.24	−.91	−.73	−.96	−.07	−1.11	−.75
Off-farm wealth	1.33*	3.65	1.17*	4.65	1.08*	3.80	1.17*	4.25
Feddans sorghum	−.04	−.41	—	—	—	—	—	—
Actual household labor[a]	—	—	.05*	−2.39	—	—	—	—
Actual labor /f cultivated	—	—	—	—	−.33*	−1.75[b]	—	—
Potential labor /f cultivated	—	—	—	—	—	—	−.22	−1.59
N	19		19		19		19	
r² adjusted	.478		.611		.558		.544	

*Significant at .05 level, one-tailed test.

[a] Actual household labor is included here but not in table A.5 because it had a nearly significant (negative) correlation with hired labor expenditures on sorghum but not on cotton.

[b] Just at significance.

TABLE A.7. Regression Equations to Predict Cotton Yields: Raw Regression Coefficients and t-ratios

Independent Variable	Equation Number					
	1		2		3	
	coef.	t-ratio	coef.	t-ratio	coef.	t-ratio
Intercept	.90	2.18	.88	1.99	.93	1.68
Hired labor Expenses/f	−.01	−.42	—	—	—	—
Feddans cotton	.09*	2.31	.08*	2.50	.08*	2.16
Actual household labor	—	—	.0001	−.12	—	—
Actual labor /f cultivated	—	—	—	—	−.02	−.20
N	19		21		21	
r² adjusted	.179		.177		.178	

*Significant at .05 level, two-tailed test.

TABLE A.8. Regression Equations to Predict Sorghum Yields: Raw Regression Coefficients and t-ratios

Independent Variable	Equation Number					
	1		2		3	
	coef.	t-ratio	coef.	t-ratio	coef.	t-ratio
Intercept	.28	.76	1.31	2.47	.92	1.64
Hired labor Expenses/f	.32*	3.71	.31*	3.54	.33*	3.49
Feddans sorghum	.06	1.45	—	—	—	—
Actual household labor	—	—	−.02	−1.57	—	—
Actual labor /f cultivated	—	—	—	—	−.07	−.64
N	19		19		19	
r² adjusted	.545		.554		.498	

*Significant at .05 level, one-tailed test.

here are discussed in chapter 7, as are the broader implications of the results.

A NOTE ON QUANTITATIVE RESULTS

Anthropologists are obligated to attempt to go beyond qualitative understandings and test ideas empirically through rigorous methods. However, quantitative analysis of anthropoligical data is difficult for a number of reasons. The lack of census and reference data and the limited resources that most fieldworkers have for the collection of quantitative data mean that sample sizes are often small. Due to the small sample size in this study, only very strong relationships are revealed. Moreover, while multiple regression equations reveal the relative strengths of independent variables in predicting a dependent variable, the effects of particular variables cannot be measured precisely in a small sample. A small sample also allows a few cases to have considerable impact on results, so outliers or exceptional cases may skew outcomes, and multiple regression equations are sensitive to outliers. Furthermore, whenever there is an attempt to quantify complex variables, the possibility of inexact correspondence between the measure and what it represents is introduced. Thus, the results of the quantitative analyses may be clouded by the lack of precise measurement of variables, such as off-farm wealth and household labor resources.

The nature of agriculture, particularly under the conditions at Wad al Abbas where rainfall and irrigation, among other things, are unpredictable, means that many other variables besides those considered here affect year to year and plot to plot labor requirements and yields.

Given these qualifications it is not surprising that multiple regressions conducted on household farming data for the 1981–82 agricultural year yielded somewhat different results than those for 1980–81. Generally, the results of these analyses confirmed the relationships between off-farm wealth, farming strategies, and yields identified in the 1980–81 data. The exception was sorghum, where off-farm wealth did not have the significant effect on hired labor use found in 1980–81 and where hired labor expenditures did not have the significant effect on yields found in 1980–81. However, none of the variables have nearly as great explanatory power in relation to sorghum in 1981–82 as they do in 1980–81. The only significant predictor of hired labor expenses on sorghum in 1981–82 was potential household labor resources/*feddan* cultivated; the relationship was negative such that households with more potential labor hired less labor. Off-farm wealth was the only significant predictor of sorghum yields that year, suggesting a productive advantage enjoyed by wealthier households. But in this case, the advantage could not be shown to be due to greater inputs of hired labor. Nor was it found to be due to their larger landholdings.

A greater number of poorer households used hired labor in 1981–82 than in 1980–81. However, the difference between the two years appears to be due to a few households who, for various reasons, are included in one sample but not the other, rather than to a change in household strategies. Expenditures on hired labor for sorghum by some poor households in 1981–82 could be explained in terms of villagers' anxieties about increasing grain prices. Villagers often remarked during this period that rising sorghum prices were leading more people to raise a sorghum crop. For the same reason—rising sorghum prices—higher investments in hired labor for sorghum would also make sense. Farmers with few resources say they are sometimes forced to sell their livestock to raise money or to sell part of their crop at harvest to pay workers back wages. In these ways, they manage to hire the labor they need. Several farmers in the sample reported that they had sold livestock or some of their sorghum to pay labor bills. Thus, on occasion, poor households may be able to raise cash for agricultural inputs through drastic means. However, they cannot do so year after year without increasing their own impoverishment. The sale of livestock that represents their source of dairy products as well as a store of wealth for emergencies and the sale of grain from household consumption supplies deplete the household's resource base.

Labor requirements of sorghum cultivation vary from year to year and from one plot to the next. Some villagers reported that there was a more favorable rain pattern for weeding in 1981–82 than the previous year, meaning lower labor requirements for sorghum that year. The accounting changes underway on the scheme may also have had some effect, perhaps altering the division of household labor inputs between cotton and sorghum. Chance risk factors, such as livestock trespass and birds, have drastic effects on sorghum yields at Wad al Abbas, reducing the effects of household strategies on yields. Therefore the effect of hired labor on productivity may be negated.

It is possible that households with greater off-farm resources have been able to secure better quality land for themselves or are more able to insure that scheme-supplied inputs, such as water, reach their land in sufficient quantity and in a timely manner. Those with greater off-farm resources, such as local traders and transporters, are more likely to have connections to scheme administration that would help them achieve this. This would help to account for the productive advantage in sorghum enjoyed by wealthier households in 1981–82. The fact that households with greater off-farm resources are more successful food producers (for whatever reasons) means that those households most dependent on farming for their survival are the least able to farm successfully.

The results of multiple regression equations reported here, based on a small sample in an unpredictable agricultural environment, should be interpreted as suggestive rather than conclusive. They do, however, lend support to conclusions reached on the basis of qualitative data.

NOTES

1. One-tailed tests are used where I have predicted the direction of the relationship, two-tailed tests where I have not. All r^2 of regression equations are adjusted for degrees of freedom; r stands for Pearsons's r correlation coefficient; t stands for the t-ratio of a variable in a regression equation. The size of a variable's t-ratio indicates how close or far the variable is from significance. The larger the t-ratio, the stronger the effect of the variable on the dependent variable and the less likely the observed relationship is the result of chance. r^2 stands for the coefficient of determination (the ratio of the explained variance to the total variance) for a regression equation; s stands for the value at which the relationship is statistically significant; N stands for the number of cases.

Glossary of Arabic Words

abid   slave.
aeed   holiday.
aesh   grain.
ahal   family.
aimma   Sudanese turban.
akhu   brother; *akhuy*, my brother.
amm   father's brother.
angaraib   traditional bed.
ankoliib   variety of sorghum.
araagi   man's garment similar to a *jellabiya* but shorter.
asel   honey; also used to refer to white fly cotton pest.
asiida   porridge (a staple of the diet).
awlaadu   his children; often used to mean a man's wife and children.
awn al zaati   self-help.
azzaaba   bachelors; bachelor style, living in a household of men.
bamia   okra.
banadura   tomatoes.
baraka   blessing, God-given powers.
bayt   house; *nas al bayt*, household, *lit.* people of the house.
bita' al hukuma   (it) belongs to the government.
bika   funeral, weeping.
bilad   privately owned rainfed farm, also called *bildat*; Villagers often refer to the time before the irrigation scheme as *al zaman al bilad* (the time of the *bilad*).
bit amm   father's brother's daughter; *pl. banaat amm.*
bit khaal   mother's brother's daughter.
bit khaalat   mother's sister's daughter.
daabit   officer.

203

dammuriya hand-woven cotton cloth; today includes coarse Sudanese machine-woven cotton.
dayn loan.
diwaan guest house or men's sitting room.
dura sorghum.
faki holy man.
fariiq neighborhood; *lit.* branch.
gaffalat al bayt she closed the house; when a woman stays with relatives during her husband's absence.
gamih wheat.
gara' gourd.
girish one piastre; *pl. groush* piastres, money.
gotiya round hut with straw roof.
gubba tomb.
gushiri an old Ethiopian coin.
hafiir water storage basin dug in the earth.
haj pilgrimage to Mecca.
hawasha holding or tenancy.
haq al hukuma (it) belongs to the government.
hisaab mushtarak joint account; also called *hisaab al ishtiraqi*; these contrast with *hisaab al ferdi* split accounts.
hosh domestic compound or courtyard.
imam leader of prayer.
jaalous traditional mud house construction.
jazira island; river island farm land.
jellaba trader from northern or central Sudan.
jellabiya man's garment, a long robe, usually white; *pl. jelaliib*.
jeref riverbank farmland; *pl. juruf*.
jibaal mountains; *s. jebel*.
jiraya long-handled hoe.
jumla en gros, wholesale.
karaama religious sacrifice.
khaal mother's brother; *khaali* my mother's brother.
khalwa Islamic school or instruction.
khariif rainy season.
khashm bayt lineage; *pl. khashum biyut*.
khudra a green leafy vegetable, *Corchorus olitorius*.
kisra Sudanese flatbread (a staple of the diet); *kisra bil moya* bread with water and usually oil and spices.
lajna board as in Board of Directors; *lajnat al intaj* production board.
ligaymaat fried doughballs.
lubia a fodder crop, *Dolichos lablab*.
lugma porridge, another word for *asiida*.
Mahdiya the period of Mahdist rule in Sudan, 1881–1898.
mahkama court of law; Mahkama Shaabi, civil court, *lit.* people's court; Mahkama Shari'a, Islamic court.
mahr bridewealth.

makhzin   storage place or warehouse.
mariisa   local beer.
markaz   center, as in *al markaz al sihhi;* health center.
masak bil nuss   sharecropping; *lit.* hold by half.
matmura   storage pit dug in the earth; *pl. mata'amir.*
mediidi   old Turkish coin.
mejlis   council.
merkoub   traditional leather shoes.
mollayta   a green leafy vegetable with a bitter taste.
mowlaana   Islamic scholar.
mowlid   the Prophet's birthday.
Mu'asessa al Ziraiyya lil Nil al Azraq   Blue Nile Agricultural Corporation.
mughterib   international labor migrant; *pl. mughteribiin.*
Mukwaar   former name of Sennar, meaning a mixture of different things.
mulaah sharmut   a stew made with dried meat.
murtaah   rested, relaxed; also means economically comfortable; *pl. murtaahiin.*
nafiir   cooperative labor party.
nasiib   male affines.
nass al barra   people outside the village.
nass al kubaar   big people or old people, respected men.
nazir   traditional leader, highest position in Native Administration.
nisba   patrilineal genealogy.
omda   traditional leader, beneath *nazir* in Native Administration; *pl. omud.*
owdat nowm   bedroom or bedroom set.
raiis   president; *pl. ruwasa.*
rakuba   thatched roof supported by poles.
rowb   curdled milk.
rushali   old Ethiopian coin.
sadd al maal   bridewealth.
saaj   metal griddle for cooking kisra.
saagia   ox-driven water-wheel for irrigation.
saif   the hot, dry season.
saiid   upstream on the Nile, places to the south.
samad   local overseer on the agricultural scheme.
sanduq   rotating credit association; *lit.* box.
sarj   local saddle, *pl. siraj.*
selluka   digging stick for farming.
semin   clarified butter.
senawi aam   junior high school, middle school.
shab'aan   sated; also means wealthy.
shari'a   Islamic law.
shariq   east; Al Shariq locally refers to the east bank of the Blue Nile.
shaykh   traditional leader, under *omda* in Native Administration; also a holy man or *faki; pl. shiyukh.*
shaykha   woman *zaar* leader; *pl. shaykhat.*
shayl   crop mortgage credit arrangement.
shita   cold, dry season.

shughul   work; *shughul al hurr*   self-employment, *lit.* free work; *shughul al hukuma*   government work.
sifinja   flip-flops; synthetic thong sandals.
simaya   baby naming ceremony.
simsin   sesame.
sirwaal   baggy pants or shorts worn under *jellabiya* or *towb*.
suq   market.
ta'ab   effort; exhaustion; *ta'baan*   tired, poor; *pl. ta'baaniin*.
tabaq   woven fiber tray cover.
ta'rifa   half a piastre.
tariiqa   Islamic brotherhood or order.
teras   traditional method of irrigation by ridging field to retain rain.
terha   girl's garment, shawl covering head to waist.
toob akhdar   unfired bricks.
towb   woman's garment, a 6-yard piece of cloth covering head to ankle.
Turkiya   Ottoman rule.
umra   Islamic pilgrimage similar to *haj*.
wad amm   father's brothers' son; *wad ammi*, my father's brother's son.
wad khaalat   mother's sister's son.
wakiil   deputy.
waggaafi   type of trader, *lit.* standing.
zaar   spirit possession cult.
zakaat   Islamic tithe.
zaawia   place used for prayer.
zikr   religious chant and dance, *lit.* remembrance.

Measures and Equivalents

CURRENCY

The Sudanese pound *(geneih)* (£S) is pegged to the U.S. dollar.
£S1 = US$1.25 in January 1980.
£S1 = US$1.11 in June 1982.
£S1 = US$0.23 in January 1988.
£S 1 = 100 piastres *(piti)* (pt.).

AREA

1 *feddan* (f) = 1.038 acres = .42 hectare.
1 *jeda'a* = 5 *feddan*s.

VOLUME

1 *guffa* (basket) = (approx.) 35 lb. unginned cotton = 15 kg.
1 *showal* (sack) = (approx.) 7.5 *kayla* = 208 lbs. = 94.6 kg sorghum.
1 small *ardeb* = 15 *kayla* = 2 sacks.
1 large *ardeb* as used at Wad al Abbas = 20 *kayla* = 2½ sacks.

WEIGHT

1 *rotl* = .99 lb. = 450 grams.
1 *kayla* = 28 *rotl* = 27.7 lb. = 12.6 kg.
1 *kantar* unginned cotton = 308 *rotl* = 312 lbs. = 141.5 kg
1 *kantar* ginned cotton = 100 *rotl* = 99.05 lb. = 45 kg.
1 metric ton = 1000 kg.

References Cited

Abdelkarim, Abbas. 1985. "The Development of Sharecropping Arrangements in Sudan Gezira: Who is Benefitting?" *Peasant Studies* 13(1)25–37.
——— 1988. "Some Aspects of Commoditisation and Transformation in Rural Sudan." In Tony Barnett and Abbas Abdelkarim, eds., *Sudan: State, Capital and Transformation*, pp. 141–160. London: Croom Helm.
Adam, Farah Hassan. 1977. "Agrarian Relations in Sudanese Agriculture: A Historical Review." *Sudan Journal of Development Research* 1(2):33–44.
Africa Recovery. 1989. Vol. 3, nos. 1–2.
Ahmed, Hassan Abdel Aziz. 1977. "Some Economic Factors Hampering the Development of Sudanese Trade During the Nineteenth Century." *Sudan Journal of Economic and Social Studies* 2(1):31–39.
Ali, Taisier Mohammed. 1983. "The Road to Jouda." *Review of African Political Economy* 23:4–14.
——— 1988. "The State and Agricultural Policy: In Quest of a Framework for Analysis of Development Strategies." In Tony Barnett and Abbas Abdelkarim, eds., *Sudan: State, Capital and Transformation*, pp. 19–36. London: Croom Helm.
Al-Shazali (Ibrahim), Salah El-Din. 1988a. "The Emergence and Expansion of the Urban Wage-Labour Market in Colonial Khartoum." In Tony Barnett and Abbas Abdelkarim, eds., *Sudan: State, Capital and Transformation*, pp. 181–202. London: Croom Helm.
——— 1988b. "The Structure and Operation of Urban Wage-Labour Markets and the Trade Unions." In Normal O'Neill and Jay O'Brien, eds., *Economy and Class in Sudan*, pp. 239–276. Brookfield: Gower.
Anthony, K., B. Johnston, W. Jones, and V. Uchendu. 1979. *Agricultural Change in Tropical Africa*. Ithaca: Cornell University Press.
Awad, Mohammed Hashim. 1971. "The Evolution of Land Ownership in the Sudan." *Middle East Journal* 25(2):212–228.

Barnett, Tony. 1977. *The Gezira Scheme: An Illusion of Development*. London: Frank Cass.
—— 1981. "Evaluating the Gezira Scheme: Black Box or Pandora's Box?" In Judith Heyer, Pepe Roberts, and Gavin Williams, eds., *Rural Development in Tropical Africa*, pp. 306–324. New York: St. Martin's Press.
Bates, Robert H. 1981. *Markets and States in Tropical Africa*. Berkeley: University of California Press.
Beer, C. W. 1955. "Social Development in the Gezira Scheme." *African Affairs* 54:42–51.
Bernal, Victoria. 1985. "Separate and Unequal: Women in a Blue Nile Village, Sudan." Paper presented at the Annual Meetings of the American Anthropological Association, Washington, D.C., December 4–8.
—— 1988a. "Losing Ground—Women and Agriculture on Sudan's Irrigated Schemes: Lessons From a Blue Nile Village." In Jean Davison, ed., *Agriculture, Women and Land: The African Experience*, pp. 131–156. Boulder: Westview Press.
—— 1988b. "Coercion and Incentives in African Agricultural Development: Insights from the Sudanese Experience." *African Studies Review* 31(2):89–108.
—— 1990. "Agricultural Development and Food Production on a Sudanese Irrigation Scheme." In Muneera Salem-Murdock and Michael M. Horowitz, eds., *Anthropology and Rural Development in North Africa and the Middle East*, pp. 197–227. Boulder: Westview Press.
Bernstein, Henry. 1988. "Capitalism and Petty-Bourgeois Production: Class Relations and Divisions of Labour." *Journal of Peasant Studies* 15(2):258–271.
Berry, Sara. 1985. *Fathers Work for Their Sons*. Berkeley: University of California Press.
Best, Alan C. G. and Harm J. de Blij. 1977. *African Survey*. New York: Wiley.
Birks, J. S. and C. A. Sinclair. 1980. *Arab Manpower*. New York: St. Martin's Press.
Bjørkelo, Anders. 1984. "Turco-*jallaba* Relations 1821–1885." In Leif O. Manger, ed., *Trade and Traders in the Sudan*, pp. 81–108. Bergen: Department of Anthropology, University of Bergen.
—— 1989. *Prelude to the Mahdiyya*. Cambridge: Cambridge University Press.
Boddy, Janice. 1989. *Wombs and Alien Spirits: Women, Men, and the Zār Cult in Northern Sudan*. Madison: University of Wisconsin Press.
Bolton, A. R. C. 1948. "Land Tenure in Agricultural Land in the Sudan." In J. D. Tothill, ed., *Agriculture in the Sudan*, pp. 187–197. London: Oxford University Press.
BNAPC (Blue Nile Agricultural Production Corporation). 1987. Annual Work Program July 1, 1986–June 30, 1987. Khartoum.
Brausch, Georges. 1964. "Change and Continuity in the Gezira Region of the Sudan." *International Social Science Journal* 16(3):341–356.
Brausch, Georges, Patrick Crooke, and John Shaw. 1964. *Bashaqra Area Settlements 1963: A Case Study on Village Development in the Gezira Scheme*. Khartoum: University of Khartoum Press.
Buttel, Frederick H. and Gilber W. Gillespie, Jr. 1984. "The Sexual Division of Farm Household Labor: An Exploratory Study of On-Farm and Off-Farm

Labor Allocation Among Farm Men and Women." *Rural Sociology* 49(2):183–209.
Chayanov, A. V. 1966. *The Theory of Peasant Economy*. Daniel Thorner, Basile Kerblay, and R. E. Smith, eds. Homewood, Ill.: Richard D. Irwin.
Chibnik, Michael. 1984. "A Cross-Cultural Examination of Chayanov's Theory." *Current Anthropology* 25(3):335–340.
Cohen, Robin, Peter C. W. Gutkind, and Phyllis Brazier, eds. 1979. *Peasants and Proletarians: The Struggles of Third World Workers*. New York: Monthly Review Press.
Collins, Carole. 1976. *Colonialism and Class Struggle in Sudan*. No. 46. Washington: Merip Reports.
Cook, Scott. 1984. "Peasant Economy, Rural Industry and Capitalist Development in the Oaxaca Valley, Mexico." *Journal of Peasant Studies* 12(1):3–40.
Cowen, Michael. 1981. "The Agrarian Problem: Notes on the Nairobi Discussion." *Review of African Political Economy* 20:57–73.
Culwick, G. M. 1955. "Social Change in Gezira." *Civilisations* 5(2):173–181.
Darling, H. S. 1951. "Insects and Grain Storage in the Sudan." *Sudan Notes and Records* 32(1):131–149.
de Janvry, Alain. 1981. *The Agrarian Question and Reformism in Latin America*. Baltimore: Johns Hopkins University Press.
Desai, Meghnad, Susanne Hoeber Rudolph, and Ashok Rudra, eds. 1984. *Agrarian Power and Agricultural Productivity in South Asia*. Berkeley: University of California Press.
Donham, Donald L. 1981. "Beyond the Domestic Mode of Production." *Man* 16(4):515–541.
——— 1985. *Work and Power in Maale, Ethiopia*. Ann Arbor: UMI Research Press.
Duffield, Mark. 1978. *Peripheral Capitalism and the Social Relations of Agricultural Production in the Village of Maiurno Near Sennar*. Economic and Social Research Council, National Council for Research, (Sudan) Bulletin No. 66.
——— 1981. *Maiurno: Capitalism and Rural Life in Sudan*. London: Ithaca Press.
——— 1983. "Change Among West African Settlers in Northern Sudan." *Review of African Political Economy* 26:45–59.
Durrenberger, E. Paul. 1984. *Chayanov, Peasants, and Economic Anthropology*. Orlando: Academic Press.
Ebrahim, M. H. S. 1983. "Irrigation Projects in Sudan." *Journal of African Studies* 10(1):2–13.
Elhassan, Abdalla Mohammed. 1988. "The Encroachment of Large Scale Mechanised Agriculture: Elements of Differentiation among the Peasantry." In Tony Barnett and Abbas Abdelkarim, eds., *Sudan: State, Capital and Transformation*, pp. 161–180. London: Croom Helm.
El Medani, Khalil Abdulla. 1986. "Macro Policies and Micro-Level Analysis of Agricultural Development in Funj Region, Sudan." Ph. D. dissertation, University of California, Riverside.
Finkler, Kaja. 1980. "Agrarian Reform and Economic Development: When Is a Landlord a Client and a Sharecropper his Patron?" In Peggy Barlett, ed., *Agricultural Decision-Making*, pp. 265–288. New York: Academic Press..

Fresco, Louise O. and Susan V. Poats. 1986. "Farming Systems Research and Extension: An Approach to Solving Food Problems in Africa." In Art Hansen and Della McMillan, eds. *Food in Sub-Saharan Africa*, pp. 305–331. Boulder: Lynne Rienner.

Gaitskell, Arthur. 1959. *Gezira: A Story of Development in the Sudan*. London: Faber and Faber.

Galal-al-Din, Mohamed El-Awad. 1988. "Sudanese Migration to the Oil-Producing Arab Countries." In Norman O'Neill and Jay O'Brien, eds., *Economy and Class in Sudan*, pp. 291–307. Brookfield: Gower Press.

Ghai, Dharam and Samir Radwan. 1983. *Agrarian Policies and Rural Poverty in Africa*. Geneva: ILO.

Gibbon, Peter and Michael Neocosmos. 1985. "Some Problems in the Political Economy of 'African Socialism.'" In Henry Bernstein, and Bonnie K. Campbell, eds., *Contradictions of Accumulation in Africa: Studies in Economy and State*, pp. 153–206. Beverly Hills: Sage.

Gleichen, A. W. E. 1905. *The Anglo-Egyptian Sudan: A Compendium Prepared by Officers of the Sudan Government*, vol. 1. London: Harrison.

Goodman, David and Michael Redclifft. 1982. *From Peasant to Proletarian, Capitalist Development and Agrarian Transitions*. New York: St. Martin's Press.

Halim, Fatimah. 1983. "The Major Mode of Surplus Labour Appropriation in the West Malaysian Countryside: The Share-Cropping System." *Journal of Peasant Studies* 10(2–3):256–278.

Hassan, Yusif Fadl. 1973. *The Arabs and the Sudan*. Khartoum: Cambridge University Press.

Herring, Ronald J. 1984. "Economic Consequences of Local Power Configurations in Rural South Asia." In Meghnad Desai, Susanne Hoeber Rudolph, and Ashok Rudra, eds., *Agrarian Power and Agricultural Productivity in South Asia*, pp. 198–249. Berkeley: University of California Press.

Hill, Polly. 1982. *Dry Grain Farming Families: Hausaland (Nigeria) and Karnatka (India) Compared*. New York: Cambridge University Press.

Hill, Richard. 1970. *On the Frontiers of Islam*. Oxford: Clarendon Press.

Holmes, Douglas R. and Jean H. Quataert. 1986. "An Approach to Modern Labor: Worker Peasantries in Historic Saxony and the Friuli Region over Three Centuries." *Comparative Studies in Society and History* 28(2):191–216.

Holt, P. M. 1969. "Four Funj Land-Charters." *Sudan Notes and Records* 50:2–14.

Holt, P. M. and M. W. Daly. 1979. *The History of the Sudan*. Boulder: Westview Press.

Horowitz, Michael M. 1989. Foreword to Muneera Salem-Murdock, *Arabs and Nubians in New Halfa*. Salt Lake City: University of Utah Press.

Hoyle, Steve. 1977. "The Khasm El Girba Agricultural Scheme: An Example of an Attempt to Settle Nomads." In Philip O'Keefe and Ben Wisner, eds., *Land Use and Development*, pp. 116–131. London: International African Institute.

Hunt, Diana. 1979. "Chayanov's Model of Peasant Household Resource Allocation." *Journal of Peasant Studies* 6(3):247–285.

Hyden, Goran. 1980. *Beyond Ujamaa in Tanzania*. Berkeley: University of California Press.

—— 1983. *No Shortcuts to Progress*. Berkeley: University of California Press.

Ibrahim, Hayder. 1979. *The Shaiqiya: The Culture and Social Change of a Northern Sudanese Riverain People*. Wiesbaden: Franz Steiner.
ILO (International Labor Organization). 1976. *Growth, Employment and Equity: A Comprehensive Strategy for the Sudan*. Geneva.
—— 1984. *Labour Markets in the Sudan*. Geneva.
—— 1986. *After the Famine: A Programme of Action to Strengthen the Survival Strategies of Affected Populations*. Geneva.
Issawi, Charles. 1966. "The Sudan." In Charles Issawi, ed., *The Economic History of the Middle East 1800–1914*, pp. 463–508. Chicago: University of Chicago Press.
Jansen, Heinz-Gerhard and Werner Koch. 1982. "The Rahad Scheme." In Gunter Heinritz, ed., *Problems of Agricultural Development in the Sudan*, pp. 23–36. Gottingen: Edition Herodot.
Jefferson, J. H. K. 1949. "The Sudan's Grain Supply." *Sudan Notes and Records* 30(1):77–98.
Kahn, Joel S. 1981. "The Social Context of Technological Change in Four Malaysian Villages: A Problem of Economic Anthropology." *Man* 16(4):542–562.
Kapteinjs, Lidwien. 1985. "Islamic Rationales for Changing the Social Roles of Women in the Western Sudan." In M. W. Daly, ed. *Modernization in the Sudan*, pp. 57–72. New York: Lilian Barber Press.
Kay, Geoffrey. 1975. *Development and Underdevelopment: A Marxist Analysis*. London: St. Martin's Press.
Khalafalla, El Fatih Shaaeldin. 1981a. "The Development of Peripheral Capitalism in Sudan: 1898–1978." Ph. D. dissertation, SUNY Buffalo.
—— 1981b. "Capital Accumulation and the Consolidation of a Bourgeois Dependent State in Sudan, 1898–1978." *Research in Political Economy* 4:29–80. JAI Press.
Kuko, Mustafa Hamza. 1984. "An Evaluation and Analysis of the Effects of Regional Economic Development on Internal Migration in the Sudan." Ph. D. dissertation, University of California, Riverside.
Laslett, Peter. 1984. "The Family as a Knot of Individual Interests." In Robert McC. Netting, Richard R. Wilk, and Eric J. Arnould, eds., *Households: Comparative and Historical Studies of the Domestic Group*, pp. 353–379. Berkeley: University of California Press.
Lehmann, David. 1986. "Two Paths of Agrarian Capitalism, or a Critique of Chayanovian Marxism." *Comparative Studies of Society and History* 28(4):601–627.
Little, Peter D. 1985. "Adding a Regional Perspective to Farming Systems Research: Concepts and Analysis." *Human Organization* 44(4):331–338.
Long, Norman. 1984. "Introduction." To Norman Long, ed., *Family and Work in Rural Societies: Perspectives on Non-Wage Labour*, pp. 1–29. London: Tavistock.
Long, Norman and Paul Richardson. 1978. "Informal Sector, Petty Commodity Production and the Social Relations of Small-Scale Enterprise." In John Clammer, ed., *The New Economic Anthropology*, pp. 176–209. New York: St. Martin's Press.

McLoughlin, P. F. M. 1962. "Economic Development and the Heritage of Slavery in the Sudan Republic." *Africa* 32(4):355–391.
——— 1970. "Labour Market Conditions and Wages in the Three Towns, 1900–1950." *Sudan Notes and Records* 51:105–118.
Mahmoud, Fatma Babikir. 1984. *The Sudanese Bourgeoisie*. London: Zed Press.
Mahmoud, Mahgoub El-Tigani. 1983. "The Impact of Partial Modernization on the Emigration of Sudanese Professionals and Skilled Workers." Ph. D. dissertation, Brown University.
Meillassoux, Claude. 1981. *Maidens, Meal, and Money*. New York: Cambridge University Press.
Millar, James R. 1970. "A Reformulation of A. N. Chayanov's Theory of the Peasant Economy." *Economic Development and Culture Change* 18:219–229.
Miskin, A. B. 1950. "Land Registration." *Sudan Notes and Records* 31(2):274–286.
Mohamed, Abdel Halim Hamid. 1984. "Resource Allocation and Enterprise Combination in a Risky Environment: A Case Study of the Gezira Scheme, Sudan." Ph. D. dissertation, Oklahoma State University.
Mohamed, Abdel Rasig El Beshir. 1986. "The Supply and Demand of Agricultural Labor." In A. B. Zahlan, and W. Y. Magar, eds., *The Agricultural Sector of Sudan: Policy and Systems Studies*, pp. 94–113. London: Ithaca Press.
Murray, Colin. 1978. "Migration, Differentiation and the Development Cycle." *African Perspectives* 1:127–143.
——— 1979. "The Work of Men, Women, and the Ancestors: Social Reproduction in the Periphery of Southern Africa." In Sandra Wallman, ed., *The Social Anthropology of Work*, pp. 337–363. London: Academic Press.
——— 1981. *Families Divided: The Impact of Migrant Labour in Lesotho*. Johannesburg: Ravan Press.
Nanda, Serena. 1984. *Cultural Anthropology*. Belmont: Wadsworth.
Niblock, Tim. 1987. *Class and Power in Sudan*. Albany: SUNY Press.
Nigam, Shyam B. L. 1977. "The Labour Requirement and Supply Situation in Agriculture in the Sudan (1973–1985)." In A. M. Al Hassan, ed., *Growth, Employment, and Equity*, pp. 140–160. A Selection of Papers Presented to the ILO Comprehensive Mission to the Sudan 1974–75. Khartoum: Khartoum University Press.
O'Brien, Jay. 1981. "Sudan: An Arab Breadbasket?" *MERIP Reports* 99:20–26.
——— 1984. "The Social Reproduction of Tenant Cultivators and Class Formation in the Gezira Scheme, Sudan." In B. Isaac, ed., *Research in Economic Anthropology* 6:217–241. Greenwich: JAI Press.
——— 1988. "The Formation and Transformation of the Agricultural Labour Force in Sudan." In Norman O'Neill and Jay O'Brien, eds., *Economy and Class in Sudan*, pp. 137–156. Brookfield: Gower Press.
O'Brien, Jay and Norman O'Neill. 1989. "Uneven Development and Class Formation." In Norman O'Neill and Jay O'Brien, eds., *Economy and Class in Sudan*, pp. 9–24. Brookfield: Gower Press.
O'Brien, John James, III. 1980. "Agricultural Labor and Development in Sudan." Ph. D. dissertation, University of Connecticut, Storrs.
Oesterdiekhof, Peter. 1980a. "Der Agrarsektor Des Sudan." In Rainer Tetzlaff and Karl Wohlmuth, eds., *Der Sudan: Probleme und Perspektiven der Entwicklung*, pp. 257–382. Frankfurt am Main: Alfred Metzner.

—— 1980b. "Agrarpolitische Orientierungen: Phasen, Tendenzen und Alternativen." In Rainer Tetzlaff and Karl Wohlmuth, eds., *Der Sudan: Probleme und Perspektiven der Entwicklung*, pp. 143–256. Frankfurt/ Main: Alfred Metzner.
O'Fahey, R. S. and J. L. Spaulding. 1974. *Kingdoms of the Sudan*. London: Methuen.
Omer, El Haj Bilal. 1985. *The Danagla Traders of Northern Sudan*. London: Ithaca Press.
O'Neill, Norman. 1978. "Imperialism and Class Struggle in Sudan." *Race and Class* 20(1):1–19
Ong, Aihwa. 1987. *Spirits of Resistance and Capitalist Discipline: Factory Women in Malaysia*. Albany: SUNY Press.
Osman, Omer Mohommed. 1958. "Some Economic Aspects of Private Pump Schemes," *Sudan Notes and Records* 39:40–58.
Osman, Omer Mohommed and A. A. Suleiman. 1969. "The Economy of Sudan." In Paul Robson, and D. A. Lury, eds., *The Economies of Africa*, pp. 436–470. Evanston: Northwestern University Press.
Parson, Jack. 1984. *Botswana: Liberal Democracy and the Labor Reserve in Southern Africa*. Boulder: Westview Press.
Randell, John R. 1958. "El Gedid: A Blue Nile Gezira Village." *Sudan Notes and Records* 39:25–39.
Reinhardt, Nola. 1988. *Our Daily Bread*. Berkeley: University of California Press.
Richards, Paul. 1985. *Indigenous Agricultural Revolution*. London: Hutchinson.
Robertson, A. F. 1987. *The Dynamics of Productive Relationships*. Cambridge: Cambridge University Press.
Rogers, Barbara. 1980. *The Domestication of Women: Discrimination in Developing Societies*. London: Tavistock.
Roseberry, William. 1976. "Peasants as Proletarians." *Critique of Anthropology* 3:3–18.
Rothstein, Frances. 1983. "The New Proletarians: Third World Reality and First World Categories." Paper presented at the Annual Meeting of the American Anthropological Association, Chicago, Ill.
Saeed, Abdel Basit. 1982. "The State and Socioeconomic Transformation in the Sudan." Ph. D. dissertation, University of Connecticut, Storrs.
—— 1988. "Merchant Capital, The State and Peasant Farmers in Southern Kordofan." In Norman O'Neill and Jay O'Brien, eds., *Economy and Class in Sudan*, pp. 186–211. Brookfield: Gower Press.
Sahlins, Marshall. 1972. *Stone Age Economics*. Chicago: Aldine.
Said, Yousif Hassan. 1968. "The Role of Agriculture in the Economic Development of Sudan." Ph. D. dissertation, University of Wisconsin.
Salem-Murdock, Muneera. 1989. *Arabs and Nubians in New Halfa: A Study of Settlement and Irrigation*. Salt Lake City: University of Utah Press.
Salih, Tayeb. 1985. *The Wedding of Zein*. Washington, D.C.: Three Continents Press.
Scott, James C. 1976. *The Moral Economy of the Peasant, Rebellion and Subsistence in Southeast Asia*. New Haven: Yale University Press.
Sellin, Harald. 1980. "Zur Entwicklung Der Handelsbourgeoisie Im Sudan." In Rainer Tetzlaff, and Karl Wohlmuth, eds., *Der Sudan—Probleme und Perspektiven der Entwicklung*, pp. 610–641. Frankfurt/Main: Alfred Metzner.

Shaner, W. W., P. F. Philipp, and W. R. Schemehl. 1982. *Farming Systems Research and Development: Guidelines for Developing Countries.* Boulder: Westview Press.
Shaw, D. J. 1966a. "The Effects of Moneylending (Sheil) on Agricultural Development in the Sudan." In D. J. Shaw, ed., *Agricultural Development in the Sudan*, pp. D56–D59. Khartoum: Philosophical Society of the Sudan.
—— 1966b. "The Development and Contribution of Irrigated Agriculture in the Sudan." In D. J. Shaw, ed., *Agricultural Development in the Sudan*, pp. 174–224. Khartoum: Philosophical Society of the Sudan.
Smith, Joan, Immanuel Wallerstein, and Hans-Dieter Evers. 1984. "Introduction." To Joan Smith, Immanuel Wallerstein, and Hans-Dieter Evers, eds., *Households and the World Economy*, pp. 7–13. Beverly Hills: Sage.
Snyder, Margaret, Nellie Okello, Mahasin Badrawi, and Khalifa Ismail. 1977. "Women and Development." In Ali Mohammed al Hassan, ed., *Growth, Employment and Equity*, pp. 232–260. A Selection of Papers presented to the ILO Comprehensive Mission to Sudan. Khartoum: Khartoum University Press.
Sørbø, Gunnar. 1977. "Nomads on the Scheme—A Study of Irrigation Agriculture and Pastoralism in Eastern Sudan." In Philip O'Keefe and Ben Wisner, eds., *Land Use and Development*, pp. 132–150. London: International African Institute.
—— 1985. *Tenants and Nomads in Eastern Sudan.* Motala: Nordiska Afrikainstitutet.
Spaulding, Jay. 1982. "The Misfortunes of Some—The Advantages of Others: Land Sales by Women in Sennar." In Margaret Jean Hay and Marcia Wright, eds, *African Women and the Law: Historical Perspectives*, pp. 3–18. Boston: Papers on Africa VII. Boston University.
—— 1984. "The Management of Exchange in Sinnar, c.1700." In Leif O. Manger, ed. *Trade and Traders in the Sudan*, pp. 24–48. Bergen: Department of Anthropology, University of Bergen.
Standing, Guy. 1981. "Migration and Modes of Exploitation: Social Origins of Immobility and Mobility." *Journal of Peasant Studies* 8(2):173–211.
Suliman, Ali Ahmed. 1975. *Issues in the Economic Development of the Sudan.* Khartoum: Khartoum University Press.
Taha, El Jack Taha. 1973. "Land Tenure and Size of Holding: Toward a New Strategy for Economic Studies of Tenancy Farming in the Gezira Scheme." *Journal of Administration Overseas* 12(4):211–218.
Tait, John. 1978. "Divisifizierung, Mechanisierung und Kapitalisierung der Produktion im Gezira Scheme. Auf dem Weg zur Uberwindung kolonial deformierter Agrarstrukturen?" *Afrika Spectrum* 13(2):165–178.
Taylor, John G. 1979. *From Modernization to Modes of Production.* Atlantic Highlands, N.J.: Humanities Press.
Voll, Sarah. 1980. "The Gezira Development Project in Sudan." In George Dalton, ed., *Research in Economic Anthropology*, 3:265–290.
Wallace, Tina. 1983. "The Kano River Project, Nigeria: The Impact of an Irrigation Scheme on Productivity and Welfare." In Judith Heyer, Pepe Roberts, and Gavin Williams, eds., *Rural Development in Tropical Africa*, pp. 281–305. New York: St. Martin's Press.
Warburg, Gabriel. 1978. "Slavery and Labour in the Anglo-Egyptian Sudan." *Journal of Asian and African Studies* 12(2):221–245.

―― 1981. "Ideological and Practical Considerations Regarding Slavery in the Mahdist State and the Anglo-Egyptian Sudan: 1881–1918." In Paul Lovejoy, ed., *The Ideology of Slavery in Africa*, pp. 245–270. Beverly Hills: Sage.

World Bank. 1988. *World Development Report*. Oxford: Oxford University Press.

Young, William Charles. 1987. "The Effect of Labor Migration on Relations of Exchange and Subordination Among the Rashaayda Bedouin of Sudan." *Research in Economic Anthropology* 9:191–220.

Yousif, Hassan Musa. 1985. "An Integrated Eco-Demographic Theoretical Framework for the Analysis of the Factors Related to the Rural Labor Force in the Gezira Scheme: A Micro Household Level Analysis." Ph. D. dissertation, University of Pennsylvania.

Index

Absenteeism, 108, 189
Abu Dhabi, employment, 75
Age; relation to agriculture, 148-50, 154-56; relation to occupation, 148 *tab*; relation to wage-work, 148-49
Agricultural labor, 47-48, 134-57, 163-76; actual in household, 168 *tab*; potential in household, 167 *tab*, 187-88; *see also* Agriculture
Agriculture: and capitalism, 3, 6-9, 27-30, 109-10, 132, 134, 154-57, 185-94; and class formation, 1, 6-9, 23-78, 181-86, 192-94; cost to household, 150-54; decline, 60-65, 68-69, 184; development policy, 186-91; food production, *see* Food; and hired labor, 31, 32-33, 35, 48, 60, 82-84, 108-9, 131, 151-52, 162-64, 168-76; and household labor, 4, 5, 26, 31, 33, 47-48, 59, 80, 82-83, 105-8, 134-57, 162-63, 166-68, 175, 187-88, 198 *tab*; impact of labor markets, 8, 134-57; income levels in, 60-64, 81, 135-38, 141, 144; inputs to, 5, 62, 162; irrigated, 29-32, 56-69, 109, 161-62 (*see also* Irrigated schemes); mechanized, 32-34; and off-farm income, 5, 7, 54, 106, 109, 134, 154-55, 162-76, 188-89; and off-farm work, 1, 9, 105, 108, 134-35, 146-50, 160-61, 181-85, 187-90; productivity, 27-38, 46, 110; relation to age, 148-50, 154-56; schedule, 44-45, 59-60; strategies, 154, 161-66; and Sudanese economy, 24; supplemented by handicrafts, 50; as survival strategy, 105-10; and trade, 27, 46, 53-56; traditional rainfed, 42-48; women's participation in, 47, 59, 91, 145, 146
Anglo-Egyptian rule, 23, 29-36, 57
Animal husbandry, 49, 56, 128-29

Black market, 65, 71, 143, 150
Blue Nile scheme, 2, 10, 57-69, 108, 189-90
Bourgeoisie: and agriculture, 26-30, 57; and government, 37; and industry, 35; and slave trade, 25; in Wad al Abbas, 16, 70-73, 78, 112, 127-30
Bridewealth, 100-1

Capital, foreign, 23, 25, 27, 29-30, 35, 37, 38

Capital accumulation, 16, 112; international, 185; and sorghum, 54; and trade, 29, 37, 53-56, 71, 73, 140-41; and transport, 37; and women, 91
Capitalism, and agriculture, see Agriculture, and capitalism
Children, and labor, 84, 95, 150-53, 158n4
Civil war, 24, 38
Class formation, 1, 6-9, 23-79, 112, 181-86, 192-94; see also Capital accumulation; Proletarianization
Consumer/producer ratios, 3
Corruption in public administration, 64
Cotton: and changes in economy, 59-60; and foreign exchange, 39; and Gezira scheme, 30; and Great Britain, 30; hired labor, 168-75, 199 tab; labor requirements, 152 tab; and migration, 32; neglect as a strategy, 135, 151-54; peasants' lack of control over, 31, 161-62; price decline, 60-61; and private management, 57-61; production schedule, 59-60; profits, 61-67, 71; rainfed production, 45; and state management, 61-69; yields, 62-67, 151-52, 201 tab
Crafts see Handicrafts
Crop-mortage system, 28, 43, 54-55, 60
Currency: introduction, 51; substitutes for, 51

Debt, foreign, 24
Development policy, agricultural, 186-91; industrial, 35; see also Irrigated schemes
Diet: changes in, 15; household hierarchy, 114
Division of labor, 83, 88-96, 145-46, 164, 167; see also headings under Agriculture)
Divorce, 100-2

Economic: changes and cotton, 59-60; cooperation between households, 85-86; cooperation within households, 85-86; inequality, 112-18, 122-30; strategies and education, 116; unit, household as, 82-88; trends, 77-78
Education: as economic strategy, 116; enrollments, 24; growth of, 74, 146-47; and income, 143-44; and occupation, 144 tab, 147 tab; and off-farm work, 115-16, 147; of women, 92-94, 117-18
Egypt, water needs, 30
Elders, household role, 104
Employment: Abu Dhabi, 75; in government, 36-37; Libya, 75; male, 88-90, 89 tab; migration for, 2; Saudi Arabia, 75; temporary, 38, 40n3; Yemen, 75; see also Wage-work
Employment figures: agriculture, 24; industry, 24; services, 24
Endogamy, 97
Ethnicity, 12-13, 84, 97
Exchange rate, 54-55; Saudi riyal, 2; Sudanese pound, 2, 111n5, 157n3, 209
Exports, 24

Family labor, see Household
Famine, 24
Farmer, see Peasant-worker class
Farming, see Agriculture
Farming systems approach, 4-5
Food: prices, 152-53, 188; production, 45-47, 188; purchase of, 81-82, 109, 154; see also Sorghum; Subsistence production
Foreign debt, 24
Foreign exchange: and cotton, 30
Foreign investment, see Capital, foreign
Funj, 25-27, 42, 43, 51

Gender: division of labor, 83, 88-96, 145-46, 164, 167; gap, 191; see also Women

Gezira scheme, 11; and cotton, 30; development, 20; income levels in, 137; management of, 2; and migrant labor, 48; and wage-labor, 33
Government: employment in, 36-37; and price controls, 71
Great Britain: and cotton, 30; in Sudan, 11, 18, 26, 29, 52, 55; *see also* Anglo-Egyptian rule

Handicrafts, 88; income levels in, 140; industry, 35; as supplement to agriculture, 49
Hired labor: agricultural, 3, 4, 5, 48, 60, 77, 108-9, 131, 161, 162; and cotton, 168-75, 199 *tab*; and credit, 63, 153; and sorghum, 168-71, 174-75, 200 *tab*; *see also* Agriculture and hired labor; Off-farm income
Household: agricultural expenses, 150-54; agricultural labor, 198 *tab* (*see also* Agriculture; and household labor); agricultural labor, actual, 168 *tab*; agricultural labor, potential, 167 *tab*; allocation of labor, 106, 162-63, 166-68; diet hierarchy, 114; economic cooperation, 85-86; as economic unit, 82-88; economy and remittances, 2, 75; farming strategies, 160-76; land use, 165-66, 197 *tab*; occupations, 106 *tab*, 121 *tab*; reproduction and off-farm work, 80-110, 182; role of elders, 104; as social unit, 86; as unit of production, 47-48, 110; use of land, 166 *tab*; wealth, 121 *tab*
Housing, 86-87

IMF, 193-94; austerity measures, 65
Income: adult, 107 *tab*; differentials, 135-42; and education, 143-44; from cotton, 60-61, 81; importance of, 76, 81; levels in agriculture, 136, 141; levels in Gezira scheme, 137; levels in handicrafts, 140; levels in off-farm work, 135-42; levels in trade, 138-39; levels in transport, 137, 139; in Nigeria, 135; occupational, 136 *tab*; per capita, 24, 37; and women, 90-94, 93 *tab*, 95, 117
Industry: development, 35; handicraft, 35; and proletarianization, 35-39
Inequality: economic, 112-18, 122-30; and landholding, 83-84, 132; and off-farm work, 112-33; and ownership of livestock, 118
Inflation, 24, 34, 38, 60, 65, 71, 150; and off-farm work, 73
Informal sector: and proletarianization, 6-7; and Wad al Abbas, 70-73, 89-91, 134, 137, 139; *see also* Trade
Inheritance practices, 29, 43-44
International Monetary Fund, *see* IMF
Investment: foreign, 37-38; *see also* Capital, foreign
Irrigated schemes: Blue Nile, 2, 10, 108; credit, 63, 152; Gezira, 2, 11, 163; individual accounts, 62-63; joint account, 58, 61-63; joint account system, 63; Khashm al Girba, 5; nationalization, 61; in Nigeria, 5; private management, 57-61; reform, 190-91; rehabilitation programs, 67; Suki, 163; state management, 61-69, 109, 161-62; Wad al Abbas, 57-69; Zeidab, 33
Islam: and crop-mortgage system, 39n1; and Zaar, 92; cultural heritage, 13; and identity, 13-14; inheritance rules, 43; and labor migration, 133n2, 143; and profit sharing, 72; and religious orders, 14
Ivory trade, 25-26, 29

Kenya, wage-work, 186
Kinship system, 17

Labor: agricultural, 134-57, 163-76; and agriculture, 134-57; by children, 84, 95, 158n4; gender division, 83, 88-96, 145-46, 164, 167; hired, 31, 48, 108, 131; household allocation, 106, 162-63, 166-68; intensification, 16, 187-88; migration, 2, 16, 35, 36, 38, 56, 69, 71-77, 90 tab, 102-4, 108 tab, 109, 116, 139-40, 142-43, 155, 190-91; migration and Gezira scheme, 32, 48; migration and household composition, 95; migration and Islam, 133n2, 143; migration and marriage, 100-1; organized, 36, 79n4, 157; requirements for cotton, 152 tab, 158n4; requirements for sorghum, 152 tab,158n4; see also Hired labor; Household

Labor markets, 3, 8, 117, 134-59

Labor migration, see Labor migration

Lancashire Cotton Growing Association, 30

Land: distribution, 83, 84 tab, 106, 132; household use, 166 tab, 197 tab; tenure, 43-44, 83, 106, 187; ownership and off-farm income, 165, 175, 196; ownership by peasant-worker class, 6-7, 35, 192-93; ownership by women, 43, 58-59; rainfed types, 43; tenancy, 57-59; value, 68-69

Language: Arabic, 23; English, 23

Levirate, 101

Libya, employment, 75

Life expectancy, 24

Livestock: damage to crops, 64, 68, 202; economic value, 49; ownership, 119n; ownership and inequality, 118; see also Animal Husbandry

Living standards, 16

Malaysia, wage-work, 186

Male employment, 88-90, 89 tab; see also Gender; Labor

Market, labor, see Labor markets

Marriage, 94-95, 97-105; delay, 100; and labor migration, 100-1; and slavery, 13, 97

Mechanized agriculture, 32-34

Merchant class, see Bourgeoisie

Mexico: landholding in, 132; off-farm work in, 161

Midwifery, 92

Migrant labor, see Labor, migration

Military service, 36

Nationalization of irrigation schemes, 61

Nigeria: incomes in, 135; irrigation schemes, 5; landholding in, 132

al-Numeiri, Jaafar, 65, 69, 100

Occupation: and education, 144 tab, 147 tab; household, 106 tab, 121 tab; and income, 136 tab; relation to age, 148 tab

Occupational specialization, 56, 72

Off-farm income: effect on agriculture, 5, 8, 54, 106, 109, 134, 154-55, 162-76, 188-89, 196; and hired labor, 3, 4, 5, 31, 154, 168-72, 170 tab; and landownership, 165, 175, 196; see also Income; Remittances

Off-farm work: and agriculture, 1-9, 105, 108-9, 134-35, 146-50, 160-61, 181-85, 187-90; and education, 115-16, 147; expansion, 69-79; and household economy, 80-110; and household reproduction, 80-110, 182; impact on agriculture, see Agriculture; income levels, 135-42; and inequality, 112-33; and inflation, 73; in Mexico, 161; and migration, 88-90; in Tanzania, 5; in trade, 70-78

Ottoman empire, 23, 25

Peasant-worker class: 1, 3-4, 6-9, 109, 110, 112, 149, 154-57, 181; and development policy, 186-91; and development theory, 4, 160-61, 181-94; formation of in Sudan, 33-39; formation of, Wad al Ab-

bas, 61-79, 104-5; future, 156, 186, 193-94; and land ownership, 7, 33, 80-81, 132
Per capita income, 24, 37
Piecework, 158n5
Political organization, 17-21
Polygyny, 10, 83, 101, 107, 111n3, 119n, 180n8
Price controls, government, 71
Productivity: agricultural, 27-38, 46, 110
Proletarianization, 1, 3, 6, 7, 8, 9, 156; and Sudanese agriculture, 32-35, 181-83; and Sudanese industry, 35-39; theories of, 6-8
Property: ownership, Wad al Abbas, 119 tab
Public administration, corruption in, 64
Purdah, see Sex segregation
Pump schemes, 30, 57; see also Irrigated schemes

Rate of exchange, see Exchange rate
Refugees, 24
Religion, see Islam
Remittances, 95, 120, 155; and household economy, 75
Residence patterns, 97-100
Riots, 60-61, 65-66, 193

Saudi Arabia and employment, 74-75, 79, 103-4, 117, 139-40
Saudi riyal, exchange rate, 2
Sennar: development, 52-53; shift of trade to, 71-72
Sex segregation, 12, 18-20, 22n3, 91
Sharecropping, 178n1, 179n7, 180n9
Shayl, see Crop mortgage
Slavery: and agriculture, 47-48; impact of, 26, 27; and marriage, 13, 97; and tenancy, 84; trade, 25, 29, 50-51
Social organization, 17, 21
Sorghum: and capital accumulation, 54; hired labor, 168-71, 174-75; household labor, 149-53; importance of, 46, 164; labor requirements, 152 tab, 158n4; necessity for household, 107; peasants' control over, 67; production and self-sufficiency, 68 tab, 122-29; production schedule, 59; yields, 174, 176, 202 tab
Sororate, 101
Specialization, occupational, 56, 72
Standards of living, 16
State: control and agriculture, 109; management and cotton, 61-67; management of agriculture, 161-62; role in capitalism, 32; see also Anglo-Egyptian rule
Subsistence production, 7, 25, 35, 42-49, 67-69, 154-57; see also Agriculture; Food; Sorghum
Sudanese pound: exchange rate, 2, 111n5, 157n3
Sudan Plantations Syndicate, 29

Tanzania, off-farm work, 5
Taxes, 34
Tenancy, 32; costs of production, 153 tab; regulations, 57-59, 62; and slaves, 84; title-holders, 111n2; transfers, 79n6, 177n1; see also Irrigated schemes; Land
Trade: and capital accumulation, 29, 37, 53-56, 71-73, 140-41; deficit, 24; income levels in, 138-39; ivory, 25-26, 29; link to agriculture, 53-56; as off-farm work, 70-78; routes, 51-53; slave, 25-26, 29, 50-51; see also Informal sector
Transport: and capital accumulation, 37; expansion, 71-72; income levels in, 137, 139
Turko-Egyptian rule, 23, 44

Unemployment, 109
Unions, labor, 36, 79n4, 157
Uxorilocal residence, 98-100, 102-4, 111n3, 180n8

Wage-work: access to, 142; and agriculture, 38, 108; and Gezira, 33; growth, 56, 69, 73-77, 190; household reproduction, 182; importance, 1; in Kenya, 186; in Malaysia, 186; and migration, 108; necessity for, 34-35; peasant dependence on, 3, 6, 8, 33, 69; relation to age, 148-49; and women, 16, 93; *see also* labor markets; Off-farm work; Peasant-worker class; Proletarianization

Wealth: distribution by household, 121 *tab*; indications of, 113-15

Women: and capital accumulation, 91; and dependence on men, 93-96; education, 92-94, 117-18; and income, 76, 90-94, 93 *tab*, 95, 117-18; land ownership, 43, 57-59; participation in agriculture, 47, 59, 91, 145, 146; and wage-work, 16, 93; *see also* Gender; Sex segregation

Work, off-farm, *see* Off-farm work

Work, wage, see Wage-work

Yemen, employment, 75

Yield: cotton, 201 *tab*; sorghum, 202 *tab*

Zaar cult, 92